T0202410

What Are the Chances of That?

WHAT ARE THE CHANCES OF THAT?

How to Think About Uncertainty

ANDREW C. A. ELLIOTT

OXFORD
UNIVERSITY PRESS

OXFORD
UNIVERSITY PRESS

Great Clarendon Street, Oxford, OX2 6DP,
United Kingdom

Oxford University Press is a department of the University of Oxford.
It furthers the University's objective of excellence in research, scholarship,
and education by publishing worldwide. Oxford is a registered trade mark of
Oxford University Press in the UK and in certain other countries

First Edition published in 2021

Impression: 1

Published in the United States of America by Oxford University Press
198 Madison Avenue, New York, NY 10016, United States of America

British Library Cataloguing in Publication Data
Data available

Library of Congress Control Number: 2021937968

ISBN 978–0–19–886902–3

DOI: 10.1093/oso/9780198869023.001.0001

Printed in Great Britain by
Bell & Bain Ltd., Glasgow

Individual–Collective

Randomness–Meaning

Foresight–Hindsight

Uniformity–Variability

Disruption–Opportunity

To Beverley, my wife.
For sharing the journey,
with all its challenges and chances.

PREFACE

Writing about uncertainty during a pandemic

The bulk of this book was written during 2019, before the outbreak of the virus that led to the Covid-19 pandemic which made 2020 such an extraordinarily difficult year. To someone writing a book about how people understand risk and uncertainty, this posed a challenge. How much of the text would have to be rewritten? I could see myself adapting examples, passages, even adding whole chapters to accommodate the disease and our response to it. On reflection, though, I decided that wholesale revision was a bad idea. If the ideas I had expressed on chance and uncertainty were right before the Covid-19 outbreak, then they were right after it, too.

Perhaps the world is now better informed about risks and more sensitive to the kinds of issues that can arise from different ways of understanding risk. But there is still little consensus on how to assess levels of risk, and the most appropriate personal and public responses. In my opinion, the pandemic has only increased the need for a better appreciation of chance in public affairs and shown how difficult clear thinking about uncertainty can be. Nowhere is this more evident than in considering the choices to be made between individual rights and collective responsibility. One of the themes in this book is the difficulty of reconciling contrasting ways of thinking about probability. Should I act in a way that best serves my own interests, exercising my own judgement and evaluation of risk, taking into account personal factors, or should I rather follow the rules, approximate and imperfect though they may be, even if I feel that my circumstances are exceptional? These issues have been highlighted by the choices we have all needed to make, and they will still be relevant when the crisis has passed. We will always have to deal with the challenges posed by uncertainty. I did not want this book to become primarily about this pandemic or for its general themes to be overshadowed by the particular issues the year 2020 has raised.

While there has been no wholesale revision of the book, Covid-19 has certainly left its mark. Where it seemed necessary to add something, I have made small changes to the text to include references to the virus in examples and in analysis. The major change is at the beginning of the chapter 'What's there to worry about?'. It seemed fitting to devote that opening to a snapshot description of the state of the outbreak in late February 2020, and my personal worries at that moment in light of the themes of the book. I have not revised that section: it can stand as an example of how we regard risks differently in prospect and in retrospect.

Peper Harow

July 2020

CONTENTS

Living in an Uncertain World

The laws of probability, so true in general, so fallacious in particular.

Edward Gibbon

Sucker bet or sure thing?

You're standing at the hotel bar. Not far along the counter is a well-dressed gentleman with a whisky in front of him who is quietly flicking through a deck of cards.

'Funny thing,' he says idly.

'What?'

'Did you know that the court cards—the king, the queen and the jack—are rather pushy characters and always turn up more often than it seems they should?'

What nonsense, you think to yourself. 'Are you sure?' you say. 'That can't be true.'

'Oh yes,' he responds, 'look here. I'll take the hearts from this pack. Thirteen cards. Ace to 10, and the jack, queen and king. I'll shuffle them. Now, you choose any three cards from these 13. If you can avoid all three court cards, you win. If jack, queen or king shows up, I win. Of course, there's a chance you might win, but I'm betting you'll lose.'

What are the Chances of That? How to think about uncertainty. Andrew C. A. Elliott,
Oxford University Press. © Andrew C. A. Elliott 2021. DOI: 10.1093/oso/9780198869023.001.0001

Three cards, you think. That should be easy. The odds are definitely with me at the start, but even after I've drawn two cards, there will still be eight non-court cards and only three court cards left. I'd still have a great chance of avoiding those, even on the third draw. Seems like it might be a good bet.

Well, do you take the bet or not? What do you reckon the chances are?

Answers at the end of this introduction.

A matter of chance; a matter of life or death

In 1999, a British solicitor, Sally Clark, was convicted of killing her two children in 1996 and 1998, at the ages of 11 weeks and 8 weeks respectively. The first child's death had been ascribed to Sudden Infant Death Syndrome (SIDS), but when a second death occurred in the family, the mother was charged, and after a long and involved trial, she was convicted.

The medical evidence presented was confusing and sometimes contradictory. One piece of non-medical evidence attracted attention. A consultant paediatrician, Sir Roy Meadow, presented as evidence a calculation that the chance of two SIDS deaths in the same family was around 1 in 73 million. He likened this to the chance of winning a bet on a rank outsider running in the Grand National horse race for four years in succession. This probability—or rather *im*probability—when presented as a compelling and dramatic comparison stood out in contrast to the confusing medical evidence, and is thought to have had considerable influence on the jury.

There was just one problem: it was wrong. The Royal Society of Statistics later prepared a report on the case, pointing out the statistical errors. When further information came to light that a forensic pathologist had withheld evidence, Sally Clark's conviction was quashed in 2003. Four years later, she died of acute alcohol poisoning.

The misleading statistical evidence had arisen from an elementary error in reasoning about probability. Meadow had taken a single statistic from a study that reported that, for a household like Sally Clark's, the prevalence of SIDS was 1 in 8,543. He interpreted this as a chance of 1 in 8,543 that a child in such a household would suffer SIDS, and to calculate the odds of this happening twice in the same household, he multiplied that figure by itself, to get to a chance of 1 in 72,982,849.

Several errors of statistical reasoning were made, but the most important is one that any student of statistics should easily have spotted. It is correct that, to reckon the combined probability of two independent events occurring, you should multiply together their individual probabilities. But take note: this rule only applies to *independent* events, where there are no plausible linking factors. Although the underlying cause or causes of the deaths of the two infants was not, and is not, known, it is strongly plausible that there was some common linking factor (or factors) in that family, or that household, that contributed to the deaths. The events were not independent, and so the calculation of the probability of two deaths as presented was incorrect. And because of the colourful way it was explained, it carried weight.

Sir Roy Meadow got his statistics wrong. Perhaps more worrying is that he was allowed to get his statistics wrong, and the error was neither detected by the judge, nor by the prosecution team of lawyers, nor even by Sally Clark's defence team. Even at the first appeal to the conviction, the statistical errors were dismissed by the appeal judges as of little significance.

Reasoning about chance is difficult. It is a subject where your intuition can easily be misled, and where there are few natural touchpoints to validate your conclusions. A definite calculation, flashily and confidently presented, carries a lot of weight. But if we want to be smarter in the way we reason about uncertainty, and avoid making errors, we need to understand where the difficulties arise.

Chance is everywhere

The true logic of this world is in the calculus of probabilities.

James Clark Maxwell

Even when we have a measurement of chance, it is not always easy to understand what the numbers mean. Some examples:

- On the eve of the 2016 US Presidential Election, the polling aggregator website *FiveThirtyEight* reported that their election model rated Donald Trump's chance of victory as around 29 per cent. At the same time, the Princeton Election Consortium had Trump's chances at around 7 per cent. Were the websites wrong? What does it mean to assign a percentage chance to a one-off event like this? After all, an election is not a repeatable event like the flipping of a coin.[1]

---

[1] In fact, running the election very many times is close to what the polling models at FiveThirtyEight.com do. Not the actual election, of course, but their constructed mathematical model of the election, including many chance elements.

- I look at the weather app on my smartphone. For 11.00 tomorrow it notes a 20 per cent chance of rain. What does this mean? Is it a 1 in 5 chance that at least one drop of rain will fall in my locality in the hour after 11.00, or does it mean that I can expect 12 minutes of rain in that hour? Or does it mean something else entirely?
- In the UK National Lottery main draw there is a 1 in 45 million chance that the numbers that come up will match my chosen six numbers. What am I to make of such a huge number? How should I understand such a small probability?
- The Covid-19 pandemic generated a flood of information about the risks involved, and what might need to be done to keep people safe. How could I tell reliable information from rumour? What was I to make of it when different authorities issued different advice?
- We hear that many species of insect are at risk of extinction. But how big is that risk? What are the consequences? How sure are the scientists? How worried should we be? And, perhaps more to the point, what actions might have a chance of addressing the problem?

We engage with uncertainty everywhere. We play childish (and not-so-childish) games that rely on chance. The stock market behaves (at least in part) as a random process. Medical diagnoses are couched in terms of probability and treatments are recommended based on statistical measures of their efficacy. Discoveries and artistic inventions happen when random influences bump up against each other.

Striving to be rational and numerate, we attempt to get a grip on this by measuring the uncertainty and assigning numbers to the chances. But, as the Sally Clark case illustrates, few non-specialists are thoroughly confident in their understanding of how chance works. Our theory of probability is relatively recent. There was no coherent way to measure uncertainty in the ancient world, and it was not until the sixteenth and seventeenth centuries that the mathematical foundations of probability were first clearly articulated. Systematic thinking about chance does not come naturally, so perhaps it should not be surprising that it still remains difficult for most of us.

For there is more to the way we think about chance than simply understanding the numbers: our feelings about uncertainty turn into hopes and fears. The statistician in me says that such emotive responses represent barriers to clear thinking. But I know—and perhaps we all know—how our own hopes and fears can distort our capacity for rational assessment. We live in a time where manipulation of sentiment is rife, and injurious to public debate. 'Fake news' deliberately plays on those hopes and fears.

Anti-vaccination campaigners, to take one example, exploit uncertainty; they amplify parents' understandable fears of doing harm to their children while downplaying the risk to the health of the population at large. Paradoxically, this increases the danger for all children. We see tensions of this kind over and over again when we look at the public presentation of uncertainty. News programmes clearly understand that the testimonies of individuals draw better audiences than the summaries of statisticians.

Modern politics is driven by fear, often based on uncertainty. A measured argument made in cautious terms, expressing a reasonable degree of doubt, is dismissed with the words 'but you can't be sure, can you?' Journalists tempt their interviewees into expressing greater certainty than is wise: 'Minister, can you guarantee that…?' These attempts to elicit absolute guarantees are in vain. The minister will either hedge, and appear weak, or they will lie, and risk being exposed when their guarantees prove to be worthless. Most of the time, complete certainty is impossible. It is better by far to develop a clear-sighted appreciation of risk and chance, and understand how they may be measured and weighed.

When an argument depends on clear facts, it should be straightforward to confirm the facts and settle the matter. But much public debate goes beyond clear facts. Predictions of rising global temperatures are probabilistic in nature. Concerns for the extinction of species are based on statistical surveys. The evidence may be strong, clear, and compelling, but few experts would claim to know absolute truth, or to be able to predict the future perfectly. In part, this is because how things turn out depends on our actions: a disaster forestalled by informed action does not mean an incorrect prediction.

Where arguments are based on likelihoods, probabilities, and predictions, there is never any chance of final certainty, and everything is a question of balanced judgement. Clear reasoning about uncertain matters is essential, and for that we need to quantify uncertainty. The scientific method always leaves room for doubt and challenge: that is part of its strength. But falling short of absolute certainty is sometimes presented as a fatal weakness. We need to be able to understand the workings of chance, how to quantify risks and opportunities, how much to trust what we hear, and so how to successfully navigate this uncertain world.

Thinking about uncertainty

I am often asked why people tend to find probability a difficult and unintuitive idea, and I reply that, after forty years researching and teaching in this area, I have finally concluded that it is because probability really is a difficult and unintuitive idea.

David Spiegelhalter

This book ranges through a variety of topics in exploring different ways that chance operates. Despite the varied contexts, though, in trying to understand why chance is so tricky to grasp, there are some patterns of thinking that keep recurring. Here's how I've organized them for this book.

Five dualities

Often, when thinking about chance, we find two points of view tugging our thoughts in different directions. These dualities (as I term them) are confusing, since the contrasting perspectives may both be valid. The tensions that are created go some way to explaining why chance can be so slippery to grasp. It's not that one way of thinking is right and the other is wrong: to think clearly about chance, you need to be aware of both perspectives, and hold them both in your mind at the same time.

Individual–Collective

Probability makes most sense when big numbers are involved. To understand the patterns that lie behind the uncertainties of life, we need data—and the more of it, the better. And yet we live our lives as individuals, as a succession of unique moments. It is hard, sometimes, to reconcile conclusions drawn from collective statistics with individual circumstances.

Randomness–Meaning

We love to make sense of what happens, and we baulk at the thought that some things happen randomly, just by chance. We want stories that explain the world, and we want ways to control what happens. Religion and science are both responses to our urge first to explain and then to control what would otherwise seem to be the meaningless operation of pure chance.

Foresight–Hindsight

Things that in prospect seemed uncertain and unpredictable can, in retrospect, appear inevitable. Hindsight is a powerful filter that distorts our understanding of chance. We are bad at prediction and very good at after-the-fact explanation: but having those explanations doesn't always mean that we will be able to avoid the next random shock.

Uniformity–Variability

The world is complex, and to understand it, we need to generalize and simplify. We often pay insufficient attention to the variety of everything around us, and are then surprised when things don't always conform to our neat, averaged expectations. We need to understand better what falls within normal ranges of variation, and which outliers genuinely merit our attention.

Disruption–Opportunity

We usually experience chance through the negative way it often affects our lives. Accidents and mistakes disrupt our plans and thwart our expectations, and this can be costly. But there are many ways in which chance can benefit us. Chance breaks stale patterns of behaviour and experience. Chance makes new opportunities, and allows new combinations. And wouldn't life be dull if all was certain?

I'll focus on each of these dualities in its own right in short sections tucked between the longer chapters of the book. And throughout the text I'll point out where one of them seems to throw particular light on the subject at hand.

What this book is not, and what it is

This is not a textbook. It won't turn you into an expert wrangler of statistics and probability. The aim is rather to develop your intuitive feel for the operation of chance. We often need to make quick decisions based on gut reactions. But when it comes to thinking about chance, our intuition is often wrong. That's why this book contains a series of explorations of the way in which chance works, the way it affects us, and the way we can use it to our benefit.

Nor is this a very mathematical book. Chance is studied by mathematicians, in the form of probability theory, but chance is also part of our lives in many other ways. Chance is embedded in our language: how we talk about it affects the way we think about it, and so I have indulged my enthusiasm for etymology and paid attention to the words we use when we talk about uncertainty.

Probability and statistics often go hand in hand, but in this book, the focus is on probability. That's not to say that I've entirely neglected statistical thinking; after all, it is an essential part of how we make sense of the randomness around us, but the weight of attention here falls on probability.

I touch on many subjects, all linked by the common factor of chance. There's a lot about gambling, but I hope you won't treat this as an invitation to gamble, still less as an instruction manual. There is discussion of chance in financial markets, but this is not a guide to investment. There's mention of disease and medical conditions, but any medical content should not be taken as authoritative. There's a section on genetics and evolution, but written from an angle that highlights the role that chance plays. I touch how computer algorithms exploit ever-growing data sets to arrive at more or less useful conclusions which are not certain, but merely probable, but this is no recipe book for machine learning.

I am treading on the turf of dozens of specialisms, and I can only hope that, in my enthusiasm to follow the trail of chance and randomness, I have not offended too many specialists. For one of the purposes of the book is precisely to go cross-country, to show that there are connected ways of thinking that disrespect boundaries and cut across the domains of finance and gambling, and genetics, and creativity, and futurism. I hope to take you to a few vantage points from which you can get a broad view of the landscape, and see how these different areas of life and knowledge are connected.

Most importantly, however serious the subject, I hope that you will find the book an entertaining read, and that it might provoke you to think about the role that chance plays in our lives, how we can strike a balance between hope and anxiety, how we can avoid the risks that are avoidable, and how we can make the most of the chances that life offers.

Getting technical

The aim of this book is to explore the subject of chance, how it affects our world and our lives, and how we think about it. I hope that almost all of the material will be easily understandable by most readers. I want to show the ways of thinking that lead to an understanding of chance, and not just a bunch of formulas and calculations. So wherever possible I will provide an intuitive and reasoned explanation rather than a technical one.

Still, there will be times when a technical explanation will provide a little something extra for the interested reader, and in this book, they will be marked by 'Getting technical' sections like this one. Feel free to skip over these technical sections: you will lose nothing essential to the arguments being made. But if you want a little more of a technical diversion, dive in!

Talking about chance

Like other occult techniques of divination, the statistical method has a private jargon deliberately contrived to obscure its methods from non-practitioners.

G. O. Ashley

It sometimes seems as if the language of chance is designed to confuse. Chances are, it is. Probability is a slippery fish. People at various times and for various purposes have tried to catch it in a net of words, and they've gone about it in different ways. The words they use to describe chance may come from different vocabularies, and because the ideas can be rather hard to grasp, they acquire a mystique. The language used becomes a protective jargon, shared by those in the know and obscure to the outsider.

A statistician's 'expected value' may not be what you expect, and their 'high level of significance' might not relate to anything that seems very significant. A financial trader will talk of 'call' and 'put' options when they might just as well say 'buy' and 'sell'. A physicist might characterize the result of an experiment in terms of 'sigmas' as a measure of confidence. In this book I will try to navigate these shoals of language, by using language that is as plain as possible and trying to explain specialist terms when we need them.

The way we choose to express probability depends on the context. If we're thinking about predicting the future, be it in regard to the weather or an impending election, it's often in terms of percentages. On the other hand, if we are thinking of populations and of how many cats prefer one brand of cat food to all others, we often talk in terms of proportions using whole numbers: 9 out of 10. If you have a gambling frame of mind, you might talk in terms of odds. A medical researcher may talk about the results of her experiment in terms of a p-value. When the particle physicists at CERN talk about how certain they are of having discovered something new, they count sigmas (standard deviations)—convention says that you need five sigmas to announce a discovery. Forecasts of future global temperatures will show widening 'funnels of doubt' that express how uncertainty increases the further into the future the projection extends. Scientists attach error bars to their published results to show their degree of confidence in the results.

Representing chances

If we want to compare how chance works across different fields, it will be helpful to use consistent methods of expressing and representing probability. Let's take a concrete example. Imagine yourself at a roulette wheel in a casino, and you bet on a single number. What are the chances that you will win on any spin of the wheel?[2] Here are some ways of showing the chances.

- Probability: 0.02632
- Percentage: 2.632%
- Proportion: 1 in 38
- Fair betting odds: 37 to 1

Sometimes it will be helpful to show this graphically:

- Proportion: 1 in 38

- Fair betting odds: 37 to 1

Another way to illustrate this is to show its effect when applied to a number of items, multiple times. For example, suppose 100 independent roulette wheels were spun 50 times. How many times would you expect your number to come up?

---

[2] Wherever roulette is mentioned in the book, I have used for my examples the so-called American wheel, which includes both a '0' and a '00'. This wheel was the original form used in Europe, and the double zero was dropped to make a more attractive game (for the players) by François and Louis Blanc, managers of a casino in Bad Homburg, in 1843. The modern European or French wheel has a single zero, which more or less halves the advantage that the casino has.

The chances are:
- For 100 wheels, over 50 spins, each with probability 2.632%, on average there will be 131.6 wins.

Then, as a simulated example:

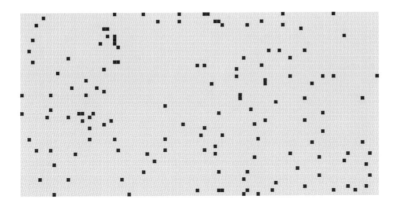

- 100 wheels × 50 spins = 5,000 chances for wins
- Observed wins in this example: 139 = 2.780% (expected average: 131.6 = 2.632%)

In the diagram above, the 100 vertical columns might represent the 100 roulette wheels, while the 50 horizontal rows represent the 50 spins for each wheel. The dark red cells represent wins on single-number bets, theoretically averaging 1 in 38. In the example simulation shown, as chance would have it, there are 139 wins, which is about 1 in 36—slightly more than average. The relative density of red cells gives a visual idea of how common or rare wins are, while the unevenness of the distribution shows how randomness can easily appear surprisingly 'clumpy'. Random does not mean evenly mixed. Note that in all these illustrations, the ratio of red to grey represents the ratio of wins to losses at the roulette wheel.

Sometimes the random events we are interested in are not ones that can be repeated, and we want to show what happens when a chance 'hit' is fatal. It's reckoned that in the Battle of Britain, each time a pilot flew a sortie, there was a 3 per cent chance that they would not make it back. Here's how we can show examples of probability of this kind, where repetition is out of the question: one hit removes everything remaining in the column. In this example, we assume that each pilot flies a maximum of 30 missions:

Representative example:

- 100 pilots × 30 sorties maximum
- Observed fatalities: 59 (expected average: 59.9)

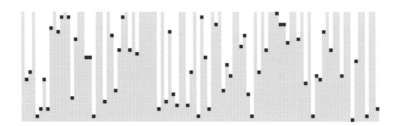

Read this diagram upwards: starting at the bottom with 100 pilots, the red cells mark the fatalities, the grey cells represent missions survived. So the grey columns that extend all the way to the top mark those 41 pilots who (in this simulation) survived all 30 missions.

Notable chances

In my book *Is That A Big Number?* I introduced the notion of a 'landmark number', a number worth remembering because it provides a mental landmark. These landmarks are useful as comparisons or mental measuring sticks to help put other numbers into context.

Understanding probability is more complex than simply grasping a single measurement, so in this book I use an adaptation of that idea. At various points in the text, I identify what I call 'notable chances', and use some of the techniques above to show the probability in numbers, words, and illustrations.

'Chance'

English offers us many words to talk about the uncertainties around us: risk, hazard, fortune, luck, peril, randomness, chaos, uncertainty. Each has its nuances and its specific connotations. For this book I have settled upon 'chance' as the central concept. It's broad in its application and evenly balanced between good and bad associations. So:

chance (n.)

> 'something that takes place, what happens, an occurrence' (good or bad, but more often bad), especially one that is unexpected, unforeseen, or beyond human control, also 'one's luck, lot, or fortune', good or bad, in a positive sense 'opportunity, favourable contingency'; also 'contingent or unexpected event, something that may or may not come about or be realised'. From
>
> > **cheance** (Old French) 'accident, chance, fortune, luck, situation, the falling of dice'. From
> >
> > > **cadentia** (Vulgar Latin) 'that which falls out', a term used in dice, from **cadere** 'to fall'. From
> > >
> > > > **kad-** (Proto-Indo-European root) 'to fall'

So I see that the language itself is guiding me to a natural starting point. For it all begins with the playing of games.

Answer to:
Sucker bet or sure thing?

You have a shuffled deck of 13 cards before you, containing all the hearts: one each of the ranks ace to 10, jack, queen, and king. For you to win this bet, three events must happen.

- First, you must select 1 card from the 13, and it must not be the jack, queen, or king.
  - The chances of this happening are 10 in 13 (roughly 76.9%)
- If that happens, then you must select 1 card from the remaining 12, and it too must not be the jack, queen, or king.
  - The chances of this happening are 9 in 12 (75%)
- If both of those happen, then you must select yet another 1 card from the remaining 11, and it too must not be the jack, queen, or king.
  - The chances of this happening are 8 in 11 (approximately 72.7%)

These events are independent of one another. To work out the chances of all three of them happening we must multiply probabilities.

Either multiply the chances expressed as fractions:

$^{10}/_{13} \times {}^{9}/_{12} \times {}^{8}/_{11} = {}^{720}/_{1716}$, which simplifies to $^{60}/_{143}$ or in decimal terms: $0.419\ldots$

or multiply the percentages:

$$76.9\% \times 75\% \times 72.7\% = 41.9\ldots\%$$

Either way, you get the same answer. Your chances of winning are less than 50 per cent. Even though the odds are with you every step of the way, when they are combined, they result in a losing situation. Don't take this bet.

PART I

Pure Chance

John Scarne: White-hat hustler

Many gamblers don't realise that the correct answers can all be obtained with figures—and that mystic symbols having to do with black cats and four-leaf clovers can be thrown out the window.

John Scarne

The iconic 1973 movie *The Sting* centres around an elaborate confidence trick perpetrated by a pair of likeable, honourable grifters (they only scam the bad guys), played by Robert Redford and Paul Newman. Of course, the charming con artists are righting a wrong: a friend has been murdered on the orders of a gangster, played by Robert Shaw, and they are determined to take revenge in the way they know best, with an elaborate confidence trick. The film is intricately plotted, and culminates in the 'sting' of the title, which springs surprise upon surprise on the audience as much as on the Shaw character.

In one memorable scene Henry Gondorff, Paul Newman's character, gets himself involved in a poker game on a train with Lonnegan, the bad guy played by Shaw. The scene includes some astonishing shots of card manipulation as Gondorff displays his dexterity. The cards seem to take on a life of their own, with the ace of spades repeatedly appearing where it could not possibly be. Paul Newman was a great actor but his skills did not extend quite so far! But neither did the card manipulations rely on some fancy special effects. Instead, the shots show the hands of magician and gambling expert John Scarne.

Scarne (1903–1985) was a showman who built his life around the skills of his hands and his ever-curious and mathematically adept brain. As a youngster, he was fascinated by seeing cardsharps performing tricks and cheats, and he trained himself to replicate their acts of dexterity. He boasted that by the age of 14 he had taught himself to stack a deck of cards while shuffling it, so he could deal a poker hand containing four aces to any seat at the card table. He might easily have been drawn into a world of gambling and cheating. Instead, persuaded by his mother to abandon the dark side, he turned his knowledge and dexterity to the good by becoming a performing magician, and then later a gambling consultant dedicated to exposing and combatting cheats in games of chance. During the Second World War he became aware of how vulnerable bored servicemen were to opportunistic cheaters, and he found himself a job on the US Army's payroll, educating the troops on how to avoid being fleeced.

In later years Scarne had an ongoing public disagreement with mathematician Edward O. Thorp about card-counting in casino blackjack. Thorp had published a book, *Beat the Dealer*, describing a technique of mentally tracking the cards that had already been dealt, which Thorp claimed would nullify the house advantage. Scarne argued that Thorp's method reduced, but did not eliminate, the dealer's edge. Despite each party challenging the other to prove their point in actual play and offers of $10,000 stakes being wagered as inducement, the reciprocal challenges were never taken up, and the argument was never resolved. Years later, though, the settled opinion has come down on Thorp's side.

Scarne was a great self-publicist, and this included a great deal of writing. His autobiography, *The Odds Against Me*, is an extremely entertaining, if somewhat self-serving, read. In 1950 he published *Scarne on Card Tricks*. For this book he collected 150 card tricks and, even though he himself was a consummate card manipulator, he reworked the tricks to remove any need for sleight of hand. The tricks are all 'self-workers', relying only on mathematics and a little psychology in the form of banter and misdirection, but not on any skills of manipulation.

Scarne knew his probability theory, and made it his work to calculate and document the probabilities and odds in gambling games of all kinds. His *Complete Guide to Gambling* runs to nearly 900 pages and is still highly regarded as a reference for the probabilities associated with the many games of chance it covers.

Pure chance

Although our intellect always longs for clarity and certainty, our nature often finds uncertainty fascinating.

Karl von Clausewitz

Tom Stoppard's play *Rosencrantz and Guildenstern Are Dead* opens with the two title characters, plucked by Stoppard from *Hamlet* and thrust into the limelight of their own play, betting on coin tosses.

Guildenstern tosses a coin, Rosencrantz calls 'heads' and wins the coin. G tosses another. R calls 'heads' and wins. Again. And again. Heads every time: 'Ninety-two coins spun consecutively have come down heads ninety-two consecutive times', says G. The characters debate what this means, including as alternatives 'divine intervention' and 'a spectacular vindication of the principle that each individual coin spun individually is as likely to come down heads as tails, and therefore should cause no surprise each individual time it does'.

We're in the audience, watching the fictitious pair on stage. We know that their world is scripted. But Stoppard is letting them in on the trick as well. He is feeding them evidence that the world they're in is not the real world, is not even pretending to be the real world, for in the real world a coin will not fall heads 92 times in a row. Without trickery, such a thing is vanishingly improbable. Only in an unreal world, a fiction, can probability be thwarted so blatantly.[3]

Flip a coin and watch it spin upwards. A thousand variables affect its brief, tumbling trajectory: the way your thumbnail catches against your finger as you flick, a stray gust of wind from an opened door, the precise geometry of its landing in your hand or on the grass. There are more than enough unmeasured factors, more than enough unpredictable variables for us to agree to call the outcome random. And the flip of the coin is just the start of it.

Perhaps the Red team called the toss correctly, and so the kick-off goes to them and not to the Blues. The boot of a Red player strikes the football—with clear intent, but not with mathematical precision—and it goes more or less, but not exactly, where the player meant it to. And now the game is alive, a game that is only worth playing or watching because we do not know what will happen, because of the chances it offers. The in-play bookmakers have set their statistical algorithms running and we are in a world of delicious uncertainty—of opportunities seized and chances missed—for 90 minutes.

Scarcely any aspect of life is untouched by unpredictability. Some phenomena have randomness at their core. The timing of the decay of a radioactive atom, most physicists agree, is in principle absolutely random, and so by definition utterly unpredictable. The coin toss, the dice roll, the spin of the roulette wheel, the dealing of the cards are things we agree to treat as random, because they are so unpredictable in practice (but perhaps not entirely so in principle). They are the outcome of complex, sometimes chaotic[4] processes involving many uncontrolled variables, and are treated *de facto* as an acceptable source of randomness because they are neither predictable, nor influenced in any exploitable way by external factors. They are not biased.

---

[3] By implication, Stoppard is telling them that everything else that is about to happen to them is also scripted. And he is telling us, the audience, 'I am about to mess with your heads.'

[4] In this context, chaotic has a precise meaning. Chaos theory involves processes that, although in principle deterministic, are in practice unpredictable.

Other kinds of uncertainty arise from ignorance at a deep level. The weather this time next week is not precisely predictable. That is not to say it is entirely random: weather systems will do what weather systems do, and follow the laws of physics in doing so, and in the short term, some broad aspects are predictable. But weather systems are extremely complex, and as we know from chaos theory, they have great sensitivity to small changes. That makes detailed prediction, beyond a certain time frame, impractical to the point of impossibility.

In this section of the book, though, we put aside the complexity factor for the most part, and we pay attention to simple situations driven by clear sources of randomness. Games of chance.

You can be fairly sure that if you were to try Rosencrantz and Guildenstern's experiment for yourself, things would go differently.[5] You'd count the heads that show up and you'd see the proportion make an erratic progress towards 50 per cent. One in two. Odds even. That's the law of large numbers: that if you perform the same experiment many, many times, the average result will home in on a particular value, the so-called expected value.

The theory of coin-flipping can be expressed as a simplified, idealized model. This abstract model, this mental construct, says that coin tosses will come up heads exactly half the time, tails exactly half the time, and never land on edge. The model does not exactly describe reality. No coin will be absolutely fair, even if its bias is undetectable in any practical sense. Still, the model is close enough, true enough, to be useful.

When we roll a cubic dice, we again have a model that guides our thinking about probability. The theoretical model of dice-rolling tells us that it will show each of 1, 2, 3, 4, 5, and 6 equally often; that is to say, one-sixth of the time. A roulette wheel has 18 red numbers, 18 black, and 2 green (0 and 00). Roulette wheel theory is that each possible number will show up on average 1 in 38 times, which means that a red can come up 18 out of 38 times, as can a black. The odds are the same for odd and even[6] numbers. One part of the theory of playing cards is that a card drawn randomly from a well-shuffled conventional 52-card deck has a 1 in 4 chance of being a diamond, a 1 in 13 chance of being an ace, and a 1 in 52 chance of being the ace of diamonds. These models of probability, these abstractions governed by well-formulated rules, are intellectual constructs that happen to give us answers that correspond closely to reality.

---

[5] This is not a provably true statement. There is of course a chance that heads will come up 92 times in a row. Sue me if it does, but make sure you have the proof!

[6] In roulette, 0 and 00 do not count as even numbers.

They are useful, and so we take them as workable approximations to truth. It's even better if the models proposed are plausible. We can reason ourselves into understanding why these formulations might work well if we can see the underlying rationale. In very broad terms, the topic of probability is about understanding how models like these create the outcomes we see, while statistics is about using what we observe in the world to generate useful models.

The world of games is a world of cosy, controlled chance, of idealized situations, of crisp statistics and neatly encapsulated theory, where the models closely correspond to what really happens. If we observe paradoxical or confusing outcomes, we can tease out the reasons, because the theories are simple and clear, the complications are few. The possible outcomes are well defined and amenable to mathematical analysis and computation. 'Random trials' can be repeated as many times as is necessary. The controlled world of a casino is one where the rules that govern chance are on display and available for study.

The casino also gives us an excellent example of one of the main principles for managing risk. For while their customers are thrill-seekers, the casino managers, who play the role of banker, are risk averse. Somewhat paradoxically, the same games when viewed from the different perspectives of player and banker look very different. They provide wild rides to the players, and steady revenue to the casino. The key to understanding this paradox lies in the numbers involved. The player will make a few dozen bets, the casino will accept many thousands.

The same logic holds true for insurance companies. They too expect to take in money steadily in smallish amounts in the form of premiums, and to make claim payments less frequently, but in much larger amounts. In the long run, the insurers, no less than the casino operators, expect to have the better side of the odds and to make money. Naturally, the service that insurers provide is not to offer thrills to their customers but to ease the pain of loss, but the principles when it comes to chance are similar. The insurance company absorbs risks, blunts the impact of those risks on policyholders, and averages out the world's misfortunes. But although the same underlying theories of probability apply in insurance and casinos, the insurance actuary's calculations have to deal with all the complexity and mess of the real world.

In his book *Black Swan*, Nassim Nicholas Taleb cautions against what he calls the ludic[7] fallacy, a seduction that tempts us to see probability in the world operating in just the same way as it does in games. Of course, this is a useful

---

[7] From the Latin *ludus*, a game.

warning: games are artificially simple, and the rules of probability as it applies in life seldom have the clarity of those we find in the casino. But artificial environments have their uses, too: if we can eliminate the myriad of external factors that muddy our thinking, we can concentrate on the bare principles involved. So for the moment, to begin to develop an intuitive feel for probability, we will stick with chance in its purest form and think about games.

What are the odds?

When it comes to gambling, people often talk in terms of odds. There are different ways of expressing chance. In a statistical context, the term most used is probability. But in the world of betting, you will probably need to understand the language of odds.

What does it mean if the odds of some event are quoted as 3 to 1? Suppose the event is whether or not a horse named Happy Jim wins the next race. The odds are a direct reference to a betting scenario. The bookmaker is offering you a possible gain of three times whatever stake you have chosen to bet (if Happy Jim wins) against the loss of that amount. A gain of three versus a loss of one. You pay over your stake when you make the bet, and should Happy Jim lose, that's it: you've lost your stake. In the event that Happy Jim wins, though, you return to the bookie with your betting slip, and he pays over your gains and also returns your original stake—you get back four times what you paid when placing the bet. Your gain, then, is three times the bet.

For odds of 1 to 1, possible gains equal possible losses. A successful bet of $10 would mean you were paid back $20 in the case of a win—your bet, plus your winnings. That's called even money, or simply 'evens'.

Traditionally, the numbers involved in odds are given as whole numbers. So odds would more usually be quoted as 5 to 2 and not 2½ to 1.[8]

If you're betting on a very likely event—a strong favourite—then the odds may be quoted as 'odds-on', where the potential gains are smaller than the losses. Odds of 1 to 2 would mean a potential gain of one for every two that you stake. Odds like this are also sometimes expressed as '2 to 1 *on*' (the numbers are reversed and the word 'on' is used).

---

[8] Historically this was the method of quoting odds. Online betting websites, where odds are calculated by computer, now often express odds as 'decimal' odds, the return always expressed in terms of a bet of 1 unit. So odds of 5 to 2 might well be expressed as 2.5 to 1 or even in terms of the overall money returned: 3.5.

When odds are used in their original context of gambling, they are really not an expression of chance at all, but simply a statement of the terms of business that the bookmaker (or casino) is offering. This is unlikely to be an exact reflection of their actual assessment of the likelihood or otherwise of the event, but will include an allowance for some profit, and may reflect other considerations, such as the need to influence the flow of bets, to balance their book, as a way of managing their overall risk.

The language of gambling (and of odds in particular) has become part of our general vocabulary. We speak of an unlikely event as having long odds. If the chance of an event is becoming greater, then the 'odds are shortening'. And when we talk about a likely event, we might say it is 'odds-on'.

It's straightforward to convert odds to an implied probability. If the odds quoted are 'odds against', take the second of the numbers quoted and divide by the sum of the two numbers. So, for our initial example, 3 to 1 odds implies a 1 in (3 + 1) chance, or a 25 per cent (0.25) probability. Odds quoted as 'evens' or 'even money' means 1 to 1. This translates into a 1 in 2 chance, 50 per cent probability.

If the odds are quoted as 'odds-on', then we're looking at a probability that is greater than 50 per cent. Use the *first* of the numbers quoted and divide by the sum of the two. So odds of '2 to 1 on' correspond to a 2 in 3 chance or a 66⅔ per cent probability.

CHAPTER 1

The Roll of the Dice

I read Shakespeare and the Bible, and I can shoot dice. That's what I call a liberal education.

Tallulah Bankhead

Another sucker bet?

You're wandering through the carnival. 'Roll up,' yells the man in the booth. 'Dice game! Three chances to win!' You take a closer look.

The man explains the rules. You pay a dollar to play, and then get to roll three normal six-sided dice. If any of the three dice shows a 6, you get your money back, plus a dollar. If a 6 shows up on two dice, you get your money back plus two dollars. And if you roll three 6s, you get your money back plus three dollars.

Now that sounds pretty good. You know that on each dice you have a 1 in 6 chance of a 6 showing. So with three dice you should be winning half the time. That gets you to break-even, right? And then, once in a while, you'll win not just $1, but $2 or $3. That's got to be a bet worth taking, hasn't it? On the other hand, though, maybe it's too good to be true?
What do you think? Scam or no?

Answers at the end of this chapter.

What are the Chances of That? How to think about uncertainty. Andrew C. A. Elliott,
Oxford University Press. © Andrew C. A. Elliott 2021. DOI: 10.1093/oso/9780198869023.003.0001

It's all in the game

On the banks of the Helmand River in south-eastern Iran, close to the border with Afghanistan, at a place called Shahr-e Sukhteh, is a Bronze Age archaeological site. The name means the Burnt City, and archaeologists reckon it is the remains of a community that arose around the year 3200 BCE. It was (as its name suggests) burned down three times and was finally abandoned in around 1800 BCE.

Discovered in 1967, the site has yielded a series of remains that paint a picture of a peaceful community (unusually, no traces of weapons have been found). Among the relics are caraway seeds, a drinking vessel with striking illustrations that some have argued constitute the earliest known animated drawing,[9] an artificial eye made of gold, and a game board. Our attention, naturally, is drawn to the game board. It represents a winding path with marked spaces, and a set of playing pieces were found along with it. All in all, it looks like a backgammon-style racing game. And crucially, along with the board and playing pieces, there was found a pair of dice, in appearance breathtakingly like modern dice.

Now, the only function of a dice[10] is to generate a random event, to allow an opportunity for chance, or luck, or fate, to show itself in the world. People choose to engage with chance for many reasons. A shaman might throw the bones looking for a message from an unseen world. A squad of soldiers might choose one of their number for a dangerous mission by drawing straws. In cricket, the choice of batting or bowling first is based on a coin toss. But if dice are involved, it's probably because a game is being played. Archaeologists have found evidence of game-playing, and specifically dice-like randomizers, all over the world.

Most games that we play for childish fun, for adult recreation, or for serious gambling incorporate chance through one mechanism or another.[11] Games are a means for us to engage with chance in a relatively innocuous way: the risks

---

[9] Each successive image has only slight differences from the previous one, suggesting frames of a movie.

[10] I was taught, of course, that the singular of dice is die. But in these modern times using die seems awkward: the singular 'dice' is now generally accepted, and that's what I use in this book.

[11] Games of skill such as chess and *Go* do not explicitly incorporate chance. These games may have formulaic openings and endgames, but in between these games are nonetheless thoroughly unpredictable, despite the absence of any random element. This reflects the complexity of the games, the vast number of possible plays, the impracticality of a complete analysis and the unpredictability of human thought. Unpredictability sometimes comes from sheer complexity.

and rewards of games are shadows of the risks and rewards we face in the uncertainties of our wider lives. Games that incorporate chance demand flexible, responsive strategies. A good player must allow not only for their opponents' actions but also for the run of luck: the roll of the dice, the fall of the cards. A little jeopardy adds excitement to our lives and games provide just a taste of it: the risk of failure, the thrill of success. This dance with uncertainty is the stuff of life, and we can easily imagine that the game-players of the Burnt City felt this as keenly as we do today.

The strength of my reaction to images of the dice found in the Burnt City startled me. Those dice seem *so* modern, so shockingly familiar. I've played dice games all my life, from *Snakes and Ladders* as a small child, through an intense obsession with backgammon at high school, an infatuation with *Dungeons and Dragons* in my university years (which introduced me to dice with 4, 8, 12, and 20 sides), and many Friday nights as an adult with my family playing intricate and engrossing modern board games. I know the feel of dice in my hand, and I know how it would feel to have those 4,000-year-old dice nestling in my palm. I know the dry, dull, clicking sound that they would make, moving against one another as I prepared to throw, and how they would tumble from the hand and roll across the table. I know that those rounded edges would make them roll a little longer, stretch by a fraction of a second the moment of suspense before they settled.

Whoever the people were who used those two dice to play their version of backgammon, I feel a kinship with them. They're my kind of people. I can imagine how they might have blown on the dice, wishing for a good roll, and how they would have cursed the times the dice would not behave themselves. They tell me that 4,000 years ago there were people like us.

Those two dice may be the oldest found, but they are far from unique. Take the time to google 'ancient dice' and look at the images that come up. Dice from China, from Egypt, from the Indus Valley, from Scandinavia. Most of them are approximately cubical, and most of them are spotted with 1 to 6 in the same formations that we see on modern dice. Some of the images you will see are of so-called long dice, stretched to the point where only four sides have a chance of coming up, the smaller end faces being unstable. You'll see teetotums and dreidels, four-sided dice designed to be spun like tops and then topple, showing one face uppermost. You will even see an Egyptian icosohedral (20-sided) dice. But the vast majority are intensely familiar: they are six-sided, and spotted from 1 to 6 in the patterns familiar to us.

Dice games and card games give us our earliest childhood exposure to pure chance. They teach us the fundamental ideas of probability: that you cannot predict the next roll of the dice, or the next flip of a card. We learn that randomness means a kind of fairness: you might bemoan your bad luck, but it's unjust to blame your opponent for it. We get to play over and over again, with the same starting point, with the same rules, and we start to learn what to expect.

What is probability anyway?

I want to plant in your mind a thought that there may be something rather strange about probability, about the very idea of measuring uncertainty, and I think that this strangeness lies at the heart of why so many of us struggle with it. For something so fundamental, it's easy to forget that for thousands of years we had no mathematical theory of chance. The modern theory of probability is relatively recent—less than 400 years old.

Imagine you're playing a game of *Snakes and Ladders* with a young person, and you have a chance to win. Your token is on the square numbered 95, and you need to land exactly on the square numbered 100. So you'll win if you roll a 5 with a single dice. What are the chances of that? Of course, anyone will tell you the chances are 1 in 6, or they might say ⅙, or then again 16⅔ per cent. Or odds of 5 to 1 against. All are ways of expressing the same idea, but let's examine that a little: what do we mean when we talk about chance in these ways?

For a start, we know that the outcome can only be one thing or another. You will either win on this roll, or you will not. Either the dice will show five spots or it will not. There is no intermediate position, no ⅙ of a win. But when I say the chance that you will roll a 5 is 1 in 6, I am really expressing a belief that if I were to roll that dice many, many times (let's say 600 times to make it concrete), then I would expect that in more or less 100 of those rolls (but probably not in *exactly* 100 of them) the number showing would be 5. So far so good. But I don't really believe that anyone has actually subjected that dice from a cheap compendium of games to any sort of rigorous testing. So my logic amounts to this: by imagining a test involving 600 dice rolls, and drawing a conclusion from that imagined test, I claim that we can say something meaningful about the dice roll you are about to make. Can that logic really be sound?

Perhaps what I mean is that if this moment in time were somehow to be replayed many, many times over, in some sort of branching-many-worlds multiverse view of reality, then in about one-sixth of those branching universes you would win the game. Or, perhaps I mean that in many, many separate games across space and time in which a situation equivalent to this one arises, where a player needs to roll a 5 to win, then in 1 in 6 of those games, they will indeed make the winning throw.

In all these interpretations, what we are saying is that the throw you are about to make is a single representative of a large number of similar situations, and we believe that on average, the number of successes will be approximately equal to a known fraction (⅙) of the total. We are claiming that we can take evidence from a large number of to-all-intents-and-purposes identical events, and find something meaningful to say about the one unique, never-to-be-repeated throw of the dice that is about to happen. That meaningful information is the highly abstract concept that we call the probability of throwing a 5.

And yet, when I roll, I will either get a 5 or not. That 1 in 6 chance will count for nothing as soon as the dice rolls to a halt with a single face uppermost. *Alea iacta est*, said Julius Caesar, before crossing the Rubicon and triggering civil war in Rome—'the die is cast'. Once rolled, the choice is made, the alternatives fall away and are meaningless.[12] Seen in prospect, six different numbers were possible. Seen in retrospect, only one of them counts.

Now you may think I am making much of something that is obvious. If you're reading this book you are likely to have at least a passing familiarity with probability. This is bread and butter to you. But I want you to look at this with fresh eyes. I am not in any way saying that the established ways of thinking about probability are wrong. But familiarity with this way of thinking may mean that we don't always see how strange an idea it is.

In the world of pure chance, in the world of rolling dice and flipping coins, we won't let this slight weirdness bother us too much. We can and will measure and wrangle the odds without too much philosophical doubt. But later, when we look at probabilities in a more complex world, where experiments are not repeatable, we'll have cause to revisit this line of thinking.

The arguments discussed so far are all based on the idea that the frequency of success is what counts, the proportion of hits to opportunities: on average, one success in every six attempts.[13] This 'frequentist' approach is not the only way to understand probability. There are other interpretations that can help us understand what probability is. You could, for example, make an argument

---

[12] Physicists, when discussing quantum phenomena, talk about the 'collapse' of the probability function, as the system falls into one particular state of the many possibilities.
[13] Misleading language alert! We are very used to making statements like 'One roll in six will be a 5.' This can so easily be misinterpreted as saying that out of every six rolls, one of them will certainly be a 5. To phrase it in that way is incorrect, but we are so used to it, we almost do not see where the error lies.

based on the nature of the dice: in particular, its symmetry. There is no discernible difference between the faces or the edges and so no reason to expect any of the numbers to be favoured over the others.[14] Happily, this argument from symmetry leads you to the same result: 1 in 6. Unhappily, there are few situations in life outside the gaming hall where the symmetries are so obligingly exact.

You might point out that the purpose of the dice is to give equal chances in a children's game. Why doubt that the manufacturer, with that aim in mind, would ensure that it was a fair game, that chances were equal? So perhaps it is reasonable to presume, even before the rolling starts, that the dice will be fair, or at least fair enough. In truth, this position is probably closest to my thinking.

And then there's the interpretation of probability in terms of a wager. When we express the chance as odds of 5 to 1 against, we are saying that if two parties with the same understanding of the matter were to make a wager, fair terms would dictate that a bet on your rolling a 5 and winning the game in that turn, should get a payout five times the stake. It reflects a consensus of opinion, a fair balance between success and failure, on terms set by mutual agreement.

These arguments are subjective: they cast probability as a measure of belief. They are all supported in one way or another by limited evidence, but they all come down to having good reason to believe in a probability of 1 in 6. Outside the casino, probabilities are very often of this kind: degrees of belief, either well or ill founded, supported by reasoning or not. We apply the rules of frequentist probability to these subjective measures, and they still make sense. Mathematicians have formalized an axiomatic theory of probability which provides a coherent underpinning for different interpretations, and there is general agreement on how to do probability calculations, but there is still debate about what probability actually is.

In this case, a subjective view of probability agrees with the frequentist logic: the chances of rolling a 5 can be taken as 1 in 6. But when it comes to matters of chance, our thinking is not always entirely logical. If you happen to be a little superstitious, you might believe that you're having a lucky streak, and feel that your chances were better than theory might indicate. You might even believe that some higher power has an interest in the game, and will take the opportunity to bless or curse your roll.

Let's step inside the casino now, that world of pure chance, but as observers, not participants. I don't plan to teach you how to gamble! We're here to watch

---

[14] This is in line with the 'propensity' theory of probability, which holds that there is an inherent tendency of a situation to behave in certain ways. This sits well with common sense, but is not easy to define rigorously.

and learn. Observing those who are gambling, we will be able to see many of the essential principles of chance at play. Let's spy on a couple of the players.

Pure chance in the casino

Sarah plays roulette

We spot Sarah standing at one of the roulette tables. Roulette is the iconic casino game: the colourful layout, the built-in suspense as the ball clickety-clacks around the spinning wheel, finally to drop into one of 38 pockets arranged around the wheel and numbered 1 to 36 plus a 0 and a 00.

For all its sparkle and colour, roulette is about as simple a gambling game as you could invent. Bets are placed on possible numbered outcomes, and the wheel is spun. If your bet covers the number that comes up, you win. If not, you lose your stake. The only complexity is that there are interesting ways to bet on multiple numbers at once, and the payouts reflect this.

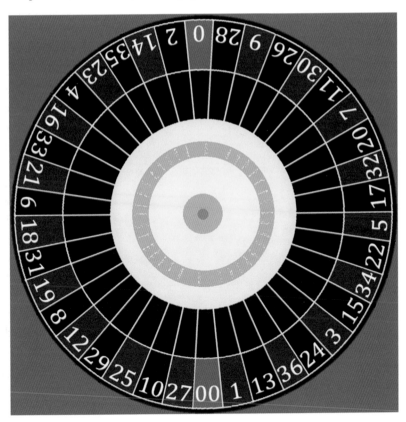

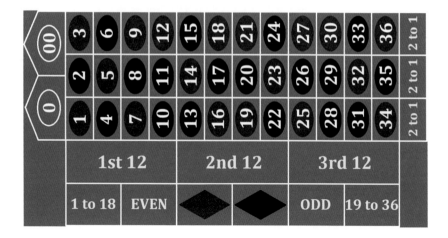

Sarah bets on 17 by placing a gambling chip squarely on the number itself, avoiding the lines that separate it from the surrounding numbers. If the ball ends up in pocket 17, Sarah wins at odds of 35 to 1. Her one chip will be returned along with 35 others. As there are 38 possible outcomes, she has a $^1/_{38}$ chance of having 36 chips returned to her; so the average return on a spin of the wheel is $^{36}/_{38}$. If Sarah were to place one chip on every number on the layout, she would certainly win on one of her bets. The croupier would gather in 38 chips and return 36. The difference of 2 is the casino's edge: around 5¼ per cent of all the bets placed.

Notable chances

In roulette the chances of winning when betting on a single number are:

- Probability: 0.02632+
- Percentage: 2.632%
- Proportion: 1 in 38

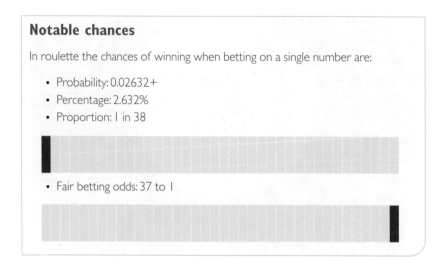

- Fair betting odds: 37 to 1

For 1,000 spins, each with probability 1 in 38, on average there will be 26.3 wins.

Actual payout on a win: 35 to 1

Margin against the player: 2 in 38 (5¼ per cent)

Representative example:

- 2,000 chances for wins
- Wins in this example: 51 = 2.70% (expected average: 52.6 = 2.63%)

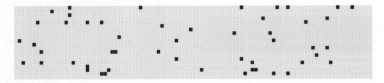

Two different things are here expressed in terms of odds. The probability of winning is 37 to 1 against, but the payout for the bet (the deal on offer) is 35 to 1. The difference is the casino's margin.

Sarah could bet on two numbers with a single chip. A chip placed on the line separating the 16 and the 17 would be taken as a split bet, and would win if either 16 or 17 came up. The payout odds in this case are correspondingly lower to reflect this greater chance of winning: 17 to 1 for such a win. These odds are consistent with the single-number bet: placing two chips on the split-bet line between 16 and 17 is equivalent to placing one on 16 and one on 17.

By placing chips on other conventional locations on the table—corners and edges—Sarah could bet on specific groups of three numbers, on four numbers, on six numbers. There are also areas for bets on sets of 12 numbers and sets of 18 numbers. The odds offered for all of these bets are mathematically consistent with the odds on a bet for a single number:

- Three-number bets pay out at 11 to 1
- Four-number bets at 8 to 1
- Six-number bets at 5 to 1
- Twelve-number bets at 2 to 1
- Eighteen-number bets at even money.

The house margin is the same on all of these: around 5¼ per cent. There is also a five-number bet covering 00, 0, 1, 2, and 3, which pays at 6 to 1. That represents worse odds for the bettor and gives the house an edge of approximately 8 per cent. One to avoid.

Notable chances

In roulette, the chances of winning when betting on red or black are:[15] (In each case, there are 18 winning numbers out of 38.)

- Probability: 0.4737
- Percentage: 47.37%
- Proportion: 9 in 19

- Fair betting odds: 10 to 9

For 1,000 spins, each with probability 18 in 38, on average there will be 473.7 wins.

Payout on a win: 1 to 1

Margin against the player: 5¼ per cent

Representative example:

- 2,000 chances to win
- Wins in this example: 971 = 48.55% (expected average: 947.4 = 47.37%)

(If you interpret this as 100 players, each playing for 20 spins, a careful examination shows that the illustration contains several runs of seven or more wins in a row—vertical streaks—and 1 of 11 wins. There are also two poor players whose sessions start with 10 losses in a row.)

Roulette systems

Generations of roulette players have tried to construct systems that produce a reliable winning strategy. As long as the equipment is functioning in an unbiased way, this is impossible. Any strategy is a combination of independent

---

[15] Or evens or odds or low numbers or high numbers.

bets on random spins of the wheel, and for every one of those spins the odds are against you. Adding together a series of losing propositions, however intricate the system, will in the long run always lose you money.

That's not to say that you can't win at roulette. Casinos would go out of business if nobody ever won! In the short term, there will always be some players who walk away as winners. It's entirely possible that Sarah will walk up to the table, place a winning bet on red, leave her winnings on the table, and see red numbers come up three times in a row. She can then walk out with eight times her initial stake.[16] But in the long run, this will happen only about 106 times in 1,000 (and her overall winnings will, on average, come to 742 times her initial stake). On the remaining occasions—about 894 times in 1,000—she will lose her initial stake (leading to losses of 894 times her stake). That makes a net loss of 152 times her stake: just over 5 per cent of the total bets she would have placed.

One system that is particularly beguiling is known as a Martingale. It's simple. Bet on one of the 18-number options that pays out at 1 to 1 (let's say you bet on red). Then if you win, all is good. Put your winnings aside, and if you wish, start the process again. But if you lose on the first spin, you must double up your stake for the next spin. If you win on the second spin, you'll win two chips, and having lost one already, you'd be one chip up overall. Put that chip aside, and start again with a single chip. However, if you lose twice in a row, you must now double your bet again; so you place four chips for the third roll. A win on the third bet will gain you four chips. Set against the three lost in the first two rolls, you are still up one chip. If you lose a third time, double up again, and continue in this way.

The logic goes like this: no matter how many times you lose, if you keep doubling the stake, then when you do eventually win, as you supposedly must do sooner or later, you will always end up one chip ahead. The logic is hard to fault. In practice, a Martingale runs into two obstacles. For a start, you would need deep pockets to keep doubling up. After just five losses you would be wagering 32 times your original stake, having lost 31 times that amount. After a run of 10 losses you would be playing with 1,024 times your initial stake. That's a big number. Secondly, casinos place limits on the size of bet they accept. Sooner or later, you will reach a point where the croupier cannot accept your bet, and you will have no way to recoup your losses.

---

[16] The idea of 'beginner's luck' may have some truth. Those who gamble for the first time and get lucky may be more likely to stick with the habit than those who lose the first time they try. So a disproportionate number of habitual gamblers may have had the experience of being lucky at the outset. It's after-the-fact reasoning, a kind of hindsight bias.

A system like the Martingale builds a complex bet out of a series of simpler bets. Yes, every time you run the Martingale sequence, you are likely to win, but your winnings will be small. And set against these numerous tiny gains, there is a small but significant chance that you might run into an irrecoverable loss. In the long run, because on each of the individual bets the odds are against you, overall the odds will be against you, too.

Let's look elsewhere in the casino. As we saw, roulette is in essence a very simple game. You place your money on one or more numbers and you win or lose. In the long run, you lose around 5¼ per cent of what you stake. But some games of chance are much more complicated than roulette.

Paul plays craps

Paul is at the craps table. In craps, the players cluster around a table designed for the throwing of dice. The player designated as the shooter rolls a pair of dice, and bets are placed on the sequence of rolls that the shooter makes. Paul looks at the dice. These dice are rather different from those found in a child's compendium of games. They have been manufactured to exacting standards, because simple probability is based on equal chances, and that calls for perfect symmetry.[17] Each face must have an equal chance of showing. The cellulose material from which these dice are made is of uniform density, and the contrasting white filler that forms the spots has the same density as the cellulose, so there is no imbalance of weight distribution that could favour one of the six sides. The dimensions are as precise as can be (a tolerance of 0.005 mm is typical), and the dice is made from transparent material to show that there is no hidden mechanism or loading weight inside. The rear of the spots may have security markings that can be inspected through the cube.[18]

For additional security, serial numbers and casino logos are visible. Additional markings may be visible only under ultraviolet light. Security protocols govern how the dice are handled by staff and players to ensure that players do not smuggle their own dice into the game. The dice are regularly tested: callipers check the dimensions and the weight distribution.

---

[17] And of course, cheating with dice can be done by using dice that are loaded with weights or have imperfect geometry. All in the interests of breaking that symmetry.

[18] Incidentally, if you are tempted to get a set of casino dice, they are not good for home play. They have sharp edges and sharp-edged cubes do not roll well on kitchen tables. In the casino, where the dice are thrown against a textured bumper backboard and land on a felted surface, they work fine.

The aim of this extravagant attention to detail is to ensure that the dice are fair, and are seen to be fair. In the case of craps, one very good reason for this is that the game is not one-sided. The bettors can bet on either the success or failure of the shooter. A bias in the dice in any direction would give the possibility of an edge against the house.

Craps is not a complex game, but with two dice and a sequence of throws for each round, it means that there is a degree of complexity around the calculation of chances. This gives plenty of scope for the casino to offer different kinds of bets, to tempt the gamblers.

There are two aspects to the game. The first relates to the mechanics of how it is played: the shooter rolls dice according to the rules and ends up either passing (think of this as hitting a target) or failing to pass (missing). Secondary to this dice-rolling done by one player at a time are the betting opportunities this presents to all of those gathered around the table. The shooter and the other players at the table place bets on the progress of the game, both before the start of the round and in the course of play. Here is a typical layout:

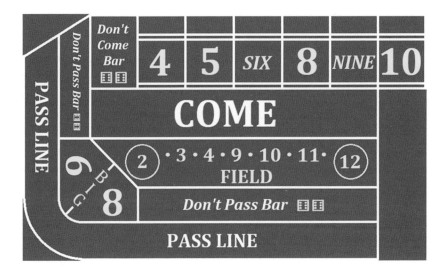

There are many opportunities to bet here, but the essence of the game relies on those spaces labelled Pass and Don't Pass. Both of those bets pay even money, 1 to 1.[19] If you win on one of those bets, you'll double your money. The question

---

[19] The Don't Pass box has a little caveat: Bar Double-Six. This is a little tweak that means that in 1 case in 36, the player neither wins nor loses on those bets. This has the effect of giving the house an overall margin.

I want to answer is: are these fair odds? How do they compare to the true probabilities of winning?

The mechanics of the game

Come-out phase:

The shooter rolls two dice in what is called the come-out roll. If the dice show a total of 7 or 11, this is a natural win. Bets on Pass win immediately. But if the dice show a total of 2, 3, or 12, this is called crapping out, and is an immediate loss: bets on Pass lose, bets on Don't Pass win.

If the total on the come-out roll is anything else, 4, 5, 6, 8, 9, or 10, the round continues. The number that was rolled is noted and is called the *point*.

Point phase:

Now the shooter's aim is to match his point. If he does this, bets on Pass will win. But if he rolls a 7 before he makes his point, that's called sevening-out, and the Don't Pass bets will be paid out. If his second roll neither hits his point, nor sevens-out, he rolls again until one of those things happens.

Craps: what are the chances?

The rules of craps are more complex than those of roulette, and calculating the chances of winning is harder, but the same approach is used. Count the ways of winning as a proportion of the number of opportunities to win.

With one dice, each number has an equal chance ($\frac{1}{6}$) of showing, but craps uses the total number of spots on two dice. Rolling a total of 7 with two dice can happen in six different ways. A roll of (1, 6) will do, as will (2, 5), (3, 4), (4, 3), (5, 2), and (6, 1). Note that (1, 6) and (6, 1) are treated as separate rolls: even though visually we can't tell the two apart, we recognize that they are different cases. And there are two ways to roll 11, namely (5, 6) and (6, 5). So taking 7 and 11 together, there are eight ways to achieve a natural win on the come-out roll.

How many different rolls in total can be made with two dice? 36. Six possibilities for the first dice combined with six possibilities for the second.[20] So 8 out of 36 possible rolls result in a natural win. These are shown as green in the diagram:

---

[20] When we look at counting the possible outcomes of combinations of two independent events, we multiply the number of possibilities of each; so in this case we calculate $6 \times 6 = 36$.

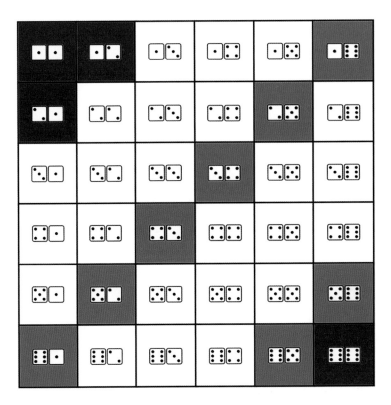

Notable chances

In craps the chances of rolling a natural win on the first roll are:

Six ways of rolling a 7: (1,6), (2,5), (3,4), (4,3), (5,2), (6,1)

Two ways of rolling an 11: (5,6), (6,5)

In total, eight ways out of a possible 36 rolls:

- Probability (rounded off): 0.2222
- Percentage (rounded off): 22.22%
- Proportion: 2 in 9

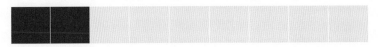

- Approximate fair betting odds: 7 to 2

continued

Notable chances *Continued*

For 1,000 come-out rolls, each with probability 8 in 36, on average there will be 222.2 natural wins.

Representative example:

- 2,000 come-out rolls
- Natural wins in this example: 469 = 23.45% (expected average: 444.4 = 22.22%)

What are the chances of crapping out?

Notable chances

In craps the chances of crapping out on the first roll are:

One way of rolling a 2: (1, 1)
Two ways of rolling a 3: (1, 2), (2, 1)
One way of rolling a 12: (6, 6)
In total, four ways out of a possible 36 rolls:

- Probability (rounded off): 0.1111
- Percentage (rounded off): 11.11%
- Proportion: 1 in 9

- Approximate fair betting odds: 8 to 1

For 1,000 come-out rolls, each with probability 4 in 36, on average there will be 111.1 crap-outs.

Representative example:

- 2,000 come-out rolls
- Crap-outs in this example: 221 = 11.05% (expected average: 222.2 = 11.11%)

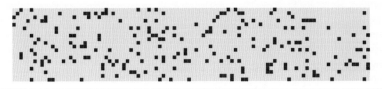

The remaining rolls (totals of 4, 5, 6, 8, 9, 10) all move the game into the point phase, and they account for 24 possibilities out of the total of 36 possible come-out rolls. So, taking all together, we have eight rolls that are naturals, four that crap out, and 24 that move along to the point phase. That totals 36. Those are all the possibilities.

Getting technical: points of principle

Probabilities are a way of talking about chance in numerical terms. Probability is always expressed as a number between 0 and 1 (or equivalently, between 0 per cent and 100 per cent). The less likely the event is, the smaller the probability of that event. Zero probability means the event is impossible.[21] A greater chance means a bigger probability, and a probability of 1 (100 per cent) means certainty. Probabilities are never less than 0 or more than 1. Outcomes that are equally likely have the same probability; so each of the 36 possible rolls of two dice has a probability of $1/36$ or 27/9 per cent, and if you add up the probabilities for all the different, mutually exclusive outcomes, the total must be 1 (100 per cent).

If two events can have no effect on one another, then they are called independent. If I eat an apple today, it has no effect on whether my doctor friend comes visiting. For events that are composed of two or more independent parts, where both or all parts must happen, the probability of the compound event is calculated by multiplying together the probabilities of the parts. If you roll two dice, there is no influence of one on the other—so we treat them as independent. To roll a double 6, each dice independently must show 6. Each has a $1/6$ probability of happening, and that means a $1/36$ probability for the compound event.

If an event can happen in two or more different mutually exclusive ways, then the overall probability is worked out by adding together the probabilities for each of the ways. There are three ways of rolling a total of 4 with two dice, namely (1, 3), (2, 2), and (3, 1). Each of these ways has a probability of 1 in 36. Add them together to get a probability of 3 in 36, or equivalently 1 in 12.

continued

[21] For practical purposes. Deep in the depths of probability theory you can encounter situations where an event with zero probability can happen. Not in the casino, though.

Craps: chances in the point phase

Let's get back to the craps game. Paul has done his come-out roll. He didn't hit a natural (7 or 11), and he didn't crap out with 2, 3, or 12. So he's now in the point phase, and his point is 10.[22] This can be rolled in three ways: (4, 6), (5, 5), or (6, 4). Of course, on the craps table (4, 6) and (6, 4) can't be told apart, but we know that they are two distinct rolls, two ways of making 10.[23] Paul needs to roll again and keep rolling until either he gets the 10 that he needs (in which case he wins, and Pass pays out) or he rolls a 7 (in which case he loses, and Don't Pass pays out).

For any roll in this phase, there are three ways for Paul to hit his 10 point and win, and six ways to hit 7 and lose. In all other cases he must roll again, and then he faces the same odds. However many rolls Paul takes, from this point on he is always twice as likely to lose as to win. So the odds that Paul ends up making his point are 2 to 1

Notable chances

Chance of making different points in the point phase:

Point	4	5	6	8	9	10
Ways to hit the point	3	4	5	5	4	3
Ways to seven-out	6	6	6	6	6	6
Proportion that hit point	3 in 9	4 in 10	5 in 11	5 in 11	4 in 10	3 in 9
Percentage	33.33%	40%	45.45%	45.45%	40%	33.33%
Odds against	2 to 1	3 to 2	6 to 5	6 to 5	3 to 2	2 to 1

[22] Many of the specific rolls in craps have nicknames. Rolling a total of 2 (1, 1) is called snake-eyes, while a total of 12 (6, 6) is box-cars. (5, 5) is puppy paws.

[23] Imagine that the two dice were two different colours: the two ways would be obvious. Historically, and to this day, this is often the root cause of misunderstanding in probability. Things can happen in two or more ways, and still look the same. It's important to count all the routes to the visible outcome.

against; that is, a 1 in 3 chance. If Paul's point had been a different number, say 9, which has four ways of coming up, his chances at this stage would be different.

If all we knew was that Paul was in the point phase, but we didn't know what his point was, what would the probability of his passing be? To calculate this is a little fiddly but not too difficult. It works out to around 41 per cent, and the chance of sevening-out is around 59 per cent. It's a weighted average of the percentages listed.

Now we know how likely it is that Paul will win on his first roll, and how likely he is to win if he needs to roll more than once. We're finally in a position to answer our main question: whether even-money odds are fair for a bet on Pass.

There are two ways that the shooter can succeed in passing:

- To pass by rolling a natural. That will happen on average eight times for every 36 come-out rolls.
  That means 22.22% of the time.
- To pass by hitting the point. For this to happen, two things are needed.
  - Firstly, the shooter must have avoided both a natural and crapping out. There is a ⅔ = 66.66% chance of this.
  - Then he must hit his point. We now know the chance of that is 40.61%.
  - We combine the two stages (by multiplying the probabilities), and conclude that winning by making the point has a 27.07% chance of happening in any round.
- Add together the probabilities for the two distinct ways of winning:

This gives a total of 49.29%. So the chance of passing is slightly less than 50%.

Notable chances

In craps the chances of winning on Pass are:

- Probability: 0.49293
- Percentage: 49.293%
- Rounded-off proportion: 35 in 71

- Approximate fair betting odds: 36 to 35

continued

Notable chances *Continued*

Actual payout on a win: 1 to 1
Margin against the player: 1 in 71 = 1.414%
Representative example:

- 2,000 rounds of craps
- Passes in this example: 989 = 49.45% (expected average: 985.9 = 49.29%)

You might notice that, although the chances here are very close to 50/50, the distribution of wins and losses has not fallen evenly. The diagram is notably patchy, and there are plenty of streaks that might persuade gamblers that the shooter had a 'hot hand'.

Since the probabilities of all outcomes must total 100 per cent, this means that the probability of Don't Pass must be 100 per cent minus 49.29 per cent, which is 50.71 per cent.

Doesn't this mean that the player betting on Don't Pass has better than even odds, an advantage over the house? Not so fast. The craps table layout has some small print. Alongside the words Don't Pass, it says Bar Double-Six. If the player craps out specifically with a double six on the come-out roll, the house will *not* pay out on Don't Pass bets. The situation will be a standoff—neither win nor lose—and the player can withdraw the bet or let it stand for the next roll. This gives the house a small margin on Don't Pass, too. Bets on Don't Pass win about 47.93 per cent of the time, and there is a standoff about 2.78 per cent of the time. This restores a slim margin (1.402 per cent) to the house.

Phew! So after all those intricate calculations, it turns out that craps is almost, but not quite, a 50/50 proposition. This may be part of what makes it a successful game: despite the complexity of the rules, it is very finely balanced between Pass and Don't Pass.

Wins on Pass are paid at odds of 1 to 1. So, on average, out of 1,000 bets, the bettor loses (and the house will get to keep the bet) 507 times (50.7 per cent). The house loses and pays out 493 times. On average, the casino profits by the equivalent of 14 bets for every 1,000 bets on Pass. On Don't Pass bets, the house pays out about 479 times out of 1,000, and will keep the bet about 493 times (and on average there will be around 28 standoffs), for a net gain of about 14 in

1,000. The house margin of 1.402 per cent on Don't Pass is one of the slimmest in the casino. (Remember that the margin for roulette is over 5 per cent.) If you want to lose your money as slowly as possible, bet on Don't Pass at craps!

For these two main bets, craps gives the casino very slim margins. But it's a fast game to play, and it can generate steady revenue. Besides, there are side bets available, where the odds are very much more in the house's favour. For example, there is a bet called 'Big 6', which will pay out at 1 to 1 if a 6 is rolled before a 7 appears. Naive gamblers may think this a fair bet, as there are three dice combinations that will make a 6—(1, 5), (2, 4), (3, 3)—and also three combinations that make a 7—(1, 6), (2, 5), (3, 4). But apart from the (3, 3), there are two different ways to make each of those other combinations. So of the 36 possible rolls with two dice, just five of them total 6, while six of them total 7, and this gives a margin of 9.1 per cent to the casino, making it something of a sucker's bet. The bet labelled 'Field' is also superficially attractive. It's a bet on the next dice roll (whatever the stage of the round) and pays out at 2 to 1 on rolls of 2 and 12, and at even money on rolls of 3, 4, 9, 10, and 11. Seven winners, and only four losing numbers (5, 6, 7, and 8). You'll not be fooled by this. You know by now that the trick lies in how many ways these totals can occur. Count them up. 16 winning ways, 20 losers. Even allowing for double payouts on 2 and 12, it is still a losing proposition.

The long run

The supernatural will continue to get a foot in at the door just as long as we try to investigate chance as it applies to a single person. But if we consider chance as it appears to a large group of players and a long series of wagers, then we begin to make sense, and superstition gets a quick brush-off.

John Scarne

Let's rejoin Sarah at the roulette table. She's betting on red. Now, there is a slightly worse than even chance that she'll win (18 out of 38 numbers are red), and a better than even chance that she'll lose (the 18 black numbers and the two green numbers 0 and 00 are losing outcomes for her). So there is a small advantage to the house, and in the long run she should expect to lose.

But there is another important asymmetry between the two sides of each bet. Sarah is there for the thrill. She is seeking risk. She doesn't know if she will walk out at the end of the evening as a winner or a loser. That risk, that hazard, is why she is there. The casino's position, on the other hand, is quite different. As with any other business, the casino's shareholders are after reliable profit. Paradoxically, they are risk *averse*. And it's the law of large numbers—a kind of statistical alchemy—that allows the casino to transmute the random and variable fortunes or misfortunes of its player-customers into a steady, stable, predictable flow of revenue.

If she keeps playing roulette, Sarah will lose in the long run. The house will win in the long run. But the house's long run will happen much sooner. Every day brings 'the long run' for the house; Sarah will probably not stay long enough for her experience to reflect the underlying probabilities. So she may end the evening with a stuffed purse, or (more likely) an emptier one, but she could not predict how much cash she would take home. However, the house management is quite confident that they will make a profit. Indeed, they'll be able to predict how much profit they should make on a given turnover. Not precisely, of course, but accurately enough for them to make plans, like any other business. And that's because they benefit from the law of large numbers.

The Gambler's Fallacy

First let's consider the Gambler's Fallacy. This is the seductive, but wrong, belief that a run of bad luck must reverse itself due to the law of averages. Surely, says the gambler, if you have flipped a coin five times and each time it has come up heads, then there must be a better than even chance of throwing tails on the sixth flip, to balance things out. No—with a fair coin the chances remain 50/50. The coin has no memory, and the previous flips have no bearing on the future. Future throws will be unbiased, no matter how streaky the past.[24] The same faulty logic has lottery players studying how long it has been since a particular ball has been selected in the draw. The lottery machine has no memory, and the history of draws has no influence on the future.

How long is the long run?

Unlike the Gambler's Fallacy, the law of large numbers can be relied on. It holds that if the same random trial (such as flipping a coin, rolling a dice, spinning a

---

[24] Of course, if the coin is in some way biased, for example unfairly weighted so as to favour heads or tails, then a streak of more heads than expected may happen. Indeed, such a streak may be the first sign of this bias. But in this chapter we are in the world of the scrupulously fair casino, and no cheating is contemplated.

roulette wheel) is performed many times, the long-term average frequency of the different possible outcomes approaches the value predicted by theory, the so-called expected value.[25] So, when flipping a coin, expect the proportion of heads to approach 50 per cent. But do not expect a streak to correct itself, still less to aim for an exact 50/50 split. Deviations from the long-run average are not corrected by future events, but rather are swamped by their numbers.

For the casino, the law of large numbers works to ensure the stability of its business. The number of bets that the house is involved in is much larger than the number of bets placed by any one bettor. In the short run, Sarah will go home as a winner or a loser. In the long run, she is almost certain to lose. But how long is the long run? On a single bet on red, Sarah's chance of winning is 18 out of 38, or about 47.4 per cent. If she bets three times, betting the same amount each time, we can calculate her chance of being a winner overall, which will happen if she wins two or three times.[26] That probability is 46.1 per cent. Her chance of being a loser overall after three bets is 53.9 per cent.

For longer betting sessions, we can go through similar calculations. I'll spare you the details, but the results look like this:

Number of bets	Chance of winning overall	Chance of losing overall
1	47.4%	52.6%
3	46.1%	53.9%
5	45.1%	54.9%
11	42.9%	57.1%
101	29.8%	70.2%
201	22.7%	77.3%
501	11.9%	88.1%
1,001	4.8%	95.2%
2,001	0.9%	99.1%

So, even if Sarah bets 201 times in the evening,[27] she still has a better than 1 in 5 (20 per cent) chance of walking out a winner. But by the time she's placed

---

[25] Probability and statistics are riddled with technical terms that are unhelpful. In this case, 'expected value' is in no sense the value that is expected from a single event.

[26] If she bets twice, she might break-even, which complicates the explanation, though not the argument. For that reason, I'll use odd numbers of bets in my examples.

[27] We're still assuming that Sarah bets only on red. If she varies her strategy, the calculations may be much more complicated, but the principles still hold.

2,001 bets, she'll probably (with more than 99 per cent certainty) have pushed her luck too far, and be a loser. On the other side of the ledger, if the casino accepts more than 1,000 bets on red on a night, they are more than 95 per cent certain to be in profit.

Notable chances

In roulette, the chances of winning overall on a succession of bets with even-money payouts are:

- 11 bets: around 43%
- 21 bets: around 40%
- 51 bets: around 35%
- 101 bets: around 30%.

Welcome to Monte Carlo

So by accepting 1,000 bets, the casino will be more than 95 per cent likely to show a profit. How much profit? Well, let's run the numbers. Imagine a very simple scenario, where each night 100 players arrive at the casino to play roulette, and each, like Sarah, places bets on red, risking the same stake (let's say $10) on each spin of the wheel. Let's assume that each player bets 11 times. This is not a particularly complex situation, and it would not be too hard to work through the theoretical mathematics of the probabilities involved. But there's another way of tackling questions like this. It's known as the Monte Carlo method. Essentially, that's just a fancy way of saying that we will set up the rules of the game in a computer (a spreadsheet will do the job perfectly well for this small simulation), and then we will 'spin the wheel' of our virtual roulette table as many times as needed using the computer's random number generator, and see how things work out.

Using a spreadsheet, I simulated one evening's play. One hundred players, 11 bets on red each, $10 a time. It turns out there were 41 overall winners out of the 100 players. The biggest winner walked away with a net gain of $90 for the evening, and the biggest loser was $70 down. Overall, the house made a profit of $420: $4.20 per player.

That was just one night. Why not simulate a few more evenings' play?

Night	Number of winners	Profit to the casino	Worst loss among the players	Best win among the players
1	41	420	70	90
2	42	660	90	70
3	38	880	90	50
4	49	560	90	90
5	47	220	70	90
6	47	300	110	70
7	40	1000	90	50
8	46	460	70	70
9	44	380	70	70
10	45	320	90	70
Average	43.9	520	84	72

Are these numbers reasonable? On each bet, the casino expects a margin of 5¼ per cent. So 1,100 bets of $10 should give an average profit near $578 per night. Here we see that the actual average ($520) was a little lower than that expected figure, but relatively close. Theory suggests a long-term average of around 42.9 winners: over the 10 nights we see a 43.9, just a little higher than theory. On average the biggest winner gained $72 from $110 staked, and the worst loser was on average $84 lighter. There is considerable variability in the experience of the players, but much less proportionate variation in the profits of the casino.

This is due to the law of large numbers. The casino was involved in 1,100 bets each night, and that gave sufficient numbers to make a difference, to dampen down the effects of randomness, and for the relatively small margin to emerge. In practice, a real-life casino will be involved in many more bets than this, and will see a bottom line that is even more free of the randomness implicit in the rules of their games.[28] For the players, though, it takes a lot of playing before the losses (that are certain in a long enough run) start to come through. There are many occasions to walk away as a winner, and the thrill of winning is a powerful reinforcement, even if the odds are always against the gamblers.

---

[28] Of course, as with any other business, the risks that casinos face will be many and various: the footfall on the night, the reliability and honesty of their staff and suppliers, the disruptive customers who cause a ruckus. Chance is everywhere, not only in the roulette wheel.

Individual experience, which makes for good stories, does not count for much by way of statistical evidence. It's difficult for the individual to extrapolate correctly from their own singular experience to general rules. If Sarah ends up as a big winner that evening, and then remembers that on her last successful visit she happened to be wearing the very same pair of shoes, it's not hard to understand why she might come to regard those as her lucky shoes. We have the idea (which is generally sound) that we can draw valid conclusions from experience, but for that process to be trustworthy, we need to understand variability, and to be wary of conclusions drawn from small amounts of data.

One way to manage risk, even profit from it, is to ensure that our experiments and ventures are numerous enough to allow risks to pool, to allow gains to offset losses, to allow the law of large numbers to operate. This can smooth away the effects of random chance. With sufficiently large numbers, we can see through uncertainty and make out the true 'rules of the game', the underlying probabilities. This is important not only because we may be interested in the workings of casinos, but because the principles apply outside the gambling house, too. Imagine an accountant employed by the casino coming across a table of data summarizing 10 evenings' play like the one shown here. He'd be able to work out just from those numbers that they probably related to even-money bets on roulette rather than Pass/Don't Pass bets from craps (which, due to the much smaller margin, would show smaller profits). That accountant, drawing inferences from studying the spreadsheet, is little different from the scientist who has performed an experiment with a hundred subjects, doing 11 trials in each of 10 batches and drawing conclusions about the underlying phenomenon. They both understand probability in the same way.

Answer to:
Another sucker bet?

It's a bad bet again, I'm afraid, though the way in which the carnie sells it is extremely persuasive, and the logic required to untangle it is intricate.

Intuition is often a bad guide in questions of chance. Three dice are rolled, and you do indeed have three chances to win your money back. But they are not three *fair* chances, regardless of how it seems. The problem lies in the paybacks for double or triple wins, which the hustler sells to you as pure bonus. In fact, they are nothing of the sort.

The trick to working this one out is to count the cases, and see what happens in each case.

When you roll three dice, there are 216 possible ways they could fall. Six ways for the first dice, multiplied by six ways for the second, multiplied by six ways for the third.

How many of those result in no payout at all? Every roll without any 6, and that means every combination of 1 to 5 on each of three dice. This can happen in 5 × 5 × 5 ways. That accounts for 125 of the 216 possibilities: more than half of them. In all of those cases, you lose your dollar.

How many rolls result in only one 6? There are 25 ways for the first dice to be 6 and the other two to show 1 to 5; there are also 25 ways for the second dice to be 6 and the other two to show 1 to 5; finally there are 25 ways for the third dice to be 6 and the other two to show 1 to 5. So there are 75 ways in total for you to win $1 in this game.

Now for the multiple wins. How many ways for two 6s to show? Five ways for there to be a 6 on dice 1 and 2 but not on dice 3; five ways for there to be a 6 on dice 1 and 3 but not on dice 2; five ways for there to be a 6 on dice 2 and 3 but not on dice 1. That makes 15 ways to win $2.

And to get three 6s? Just the one way.

So, bring it all together: of the 216 possible rolls:

125 will lose you a dollar
75 will win you a dollar
15 will win you two dollars
1 will win you three dollars.

If you played 216 times, and every possible combination came up, your losing rolls would cost you $125. Your winning rolls would gain you $1 × 75 + $2 × 15a $3 × 1 = $106. Overall, you would be down $19.

Why? Well the obvious, but false, logic says that this is just like having three bets, one on each of the dice. But when you bet on some proposition and win, you get back your original stake. If you really did have a $1 bet on each dice, you'd get paid $2 for every 6 that came up (your stake plus your winnings). So to make this game fair, you need a bonus of $2 for every extra 6 that came up. Two 6s? The fair winnings should be $3. Three 6s? The winnings should be $5. That's the logic, but let's work through the cases to check:

Under the new rules, which we think will be fair, of the 216 possible rolls:

125 will lose you a dollar
75 will win you a dollar

15 will win you three dollars

1 will win you five dollars.

If you played 216 times and every possibility came up once, your losing rolls would still cost you \$125. But your winning rolls would gain you $\$1 \times 75 + \$3 \times 15 + \$5 \times 1 = \$125$.

Now *that* would be a fair game.

The first duality: Individual–Collective

The National Lottery was launched in the United Kingdom in 1994 with the slogan 'It could be you'. A smart move. In cold statistical terms, the chance of winning the main prize was a rather off-putting 1 in 14 million. But, by showing a huge hand descending from the sky and pointing out Joe Soap sitting on a park bench, they made it personal. For, after all, it *could* be you. And it's all too tempting when faced with a risk such as being infected by a virus, which has low probability but potentially serious consequences, to think 'it's not going to happen to me'.

Flip a coin once. The outcome is heads or tails and there's not much to be learned from that. Flip it 10 times: if it's a fair coin, there is less than a 1 in 1,000 chance that they are all heads. Flip the coin 10,000 times, keeping count of heads and tails, and you might have enough information to decide whether or not, for *that* coin, the chances are indeed 50/50. This is the alchemy of large numbers: valuable information can be extracted from base randomness. With enough observations, we can detect any signals that may be hiding in the noise of random experience, and random experiment. But don't fool yourself. Even if you understand the odds of throwing 10 heads in a row, when it comes to the 11th flip, you can't guess which side it will land.

I am an individual. I move through the world in a single time-line, experiencing unique events. Though I share much with others, my personal experience will always be mine alone. Yet at the same time, I am one of many who are like me in many ways. I can be tagged, classified, and analysed as a statistic. Mortality tables can predict with some accuracy how many people like me are expected to die at my age, even though every death will be different.

Reasoning from individual instances is seldom a good basis for deriving general rules. To draw conclusions, you need more data, and the more you have, the better the chance that you will get a glimpse of what lies behind the randomness. On the other hand, even if you know the odds, this is no sure guide to a safe decision in individual cases. I may agree to a medical procedure knowing it has a small risk of going badly. But if things go wrong, it matters not a bit that the odds were in my favour.

So this is our first duality: when it comes to chance, numbers matter. Thinking collectively does not always deliver the best outcome for each individual. And nor does each person choosing for themselves always result in the best outcome for the group.

CHAPTER 2

Reckoning the Chances

We all that are engagéd to this loss
Knew that we ventured on such dangerous seas
That if we wrought out life 'twas ten to one

William Shakespeare, Henry IV, Part II

Count the ways, and then count again

The great Galileo was posed a conundrum by a baffled gambler. Roll three dice. How many combinations make a total of 9? The answer, the gambler suggested, was 6:

1 + 2 + 6, or
1 + 3 + 5, or
1 + 4 + 4, or
2 + 2 + 5, or
2 + 3 + 4, or
3 + 3 + 3.

How many combinations make a total of 10? The answer, once again, the questioner claimed, is 6:

1 + 3 + 6, or
1 + 4 + 5, or
2 + 2 + 6, or
2 + 3 + 5, or
2 + 4 + 4, or
3 + 3 + 4.

What are the Chances of That? How to think about uncertainty. Andrew C. A. Elliott,
Oxford University Press. © Andrew C. A. Elliott 2021. DOI: 10.1093/oso/9780198869023.003.0002

How then, could it be the case (as everyone knew) that 10 came up marginally more often than 9?

Answer at the end of this chapter.

How will the bones roll?

The ancient Greek historian Herodotus, 'the father of history', had it that dice games were invented by the Lydians around 3,000 years ago:

> In the reign of Atys son of Manes there was great scarcity of food in all Lydia. For a while the Lydians bore this with what patience they could; presently, when the famine did not abate, they looked for remedies, and different plans were devised by different men. Then it was that they invented the games of dice and knuckle-bones and ball and all other forms of game except draughts, which the Lydians do not claim to have discovered.

So dice games were invented as a distraction from hunger. Perhaps not: Herodotus is also known as the father of lies. And dice are found in the archaeological record dating from long before the time of which Herodotus writes. Even older than the remarkable stone dice found in Iran's Burnt City are the great numbers of animal knucklebones that have been found in archaeological sites dating back thousands of years. Cattle, goats, sheep, and deer have a bone known as the *astragalus* at the location in the hind leg that corresponds to the ankle in a human.[29] This bone, small enough to hold in your hand, has four flattish surfaces, and when thrown onto a playing surface, will land in one of four stable positions, showing one of the faces uppermost.

These bones have been found around the world, with clear indications that they were used in games and gambling. That they were used as randomizers is confirmed by the presence in some cases of clear distinguishing marks on the four sides that might show uppermost. That they were handled repeatedly is shown by distinctive patterns of wear. Knucklebones have been found together with game pieces used for the ancient Egyptian game of Senet, and it is thought that *astragali* were used for game-playing in Egypt as early as 3500 BCE. As to the rules of the games that the ancient Egyptians played, we have little clear evidence.

When it comes to the Romans, though, we know considerably more. The game of *Tali* seems to have involved throwing four *astragali* at a time, and using

---

[29] So, not from the knuckle at all…

the various combinations that resulted to signal specific actions to be carried out by the players. For *tali*, the Romans assigned point values of 1, 3, 4, and 6 to the four faces—so they were clearly aware of the different chances for each. The lowest-ranked face, with a point value of 1, was called the dog. A throw that consisted of four dogs was also called a dog, while a combination where the four *astragali* each showed a different face was called a Venus, and was regarded very favourably.

A knucklebone is not perfectly symmetrical—so the chances of each orientation are not equal. The statistician Florence David[30] estimated, based on her own trials, that for a sheep's *astragalus* the flatter sides had a 4 in 10 chance of showing, the narrower faces a 1 in 10 chance. But although the knucklebone's four sides have different chances, nowhere has there been found any evidence that these differences were quantified by the Romans in any way. Through games of chance, people have engaged actively with what we now call probability for millennia, but for most of that time there was no way of getting a numerical grip on uncertainty—no way of calculating chances.

The question of why this branch of mathematics lay unexplored for so many centuries is an intriguing one, and may have some bearing on why even now, calculations of probability are so liable to trip us up. It's an area where our instincts do not serve us well.

'All is number', said Pythagoras, but for the ancient Greeks, that did not extend as far as measuring probability. Aristotle took a qualitative approach. In the classification he adopted, events could be certain, probable, or unknowable, and games of chance fell into the unknowable category. If the great Aristotle had ruled out the possibility of knowledge of such things, who would waste their time exploring that subject? Religious belief might also have played a role in discouraging a study of chance. If the outcome of uncertain events was due to the intervention of an unseen power, be it a personified deity or abstract fate, then to presume to make calculations of probability was hubristically to seek to second-guess the mind of God (or the gods).

But we are intellectually curious, and stubborn, and not a little greedy, and there is little that will dissuade an enquiring mind from organizing knowledge and ideas, if there can be a hope of profiting commercially from better decisions and more favourable outcomes.

---

[30] Her full name: Florence Nightingale David.

Risky voyages

Financial self-interest is a powerful motivator. When a deal is made, a price is needed. And, if that deal includes indemnifying a risk, then a fair price for the risk is needed. In ancient Athens, a trader proposing a voyage would seek a maritime loan, one not repayable in the event of the ship being lost. The return due to the lender (in the event of the safe completion of the trip) was increased to reflect the degree of risk involved, the amount of the increase depending on the season of the year. So the agreement as a whole incorporated a form of embedded insurance, with an implied price for covering the risk, although the risk element was never explicitly recognized.

In medieval Europe, by the thirteenth century, similar arrangements were available for merchant ventures. Sea voyages for the purposes of trading were perilous but, if perils were avoided, they could be very profitable. Contracts were struck to fund such voyages—contracts that explicitly recognized the risk of wreck and consequent loss of any investment—and the terms of the deal made specific allowance for the degree of danger anticipated. This certainly represents an attempt to place a numerical value on the risk, although it seems never to have been expressed in these terms. Rather, the price struck was the consequence of a negotiation between knowledgeable parties, who agreed that the terms represented a fair deal in consideration of the risk involved. Such a price was a quantification of a subjective opinion, and was reached by bargaining, much as the price of a horse or a house might have been.

The earliest stand-alone insurance policy[31] that we know of dates from 1347 and it was written in Genoa, although it appears to document a well-established practice. Still, premiums for such policies were not mathematically based, but were, as in earlier times, based on informed opinion as to the risk, and agreed in negotiation.

Quantifying chance

As we see from the opening quotation, William Shakespeare, writing in the late sixteenth century, was confident that his audience would be familiar with the terminology of odds: it's the fundamental language of betting. But, while gambling people and insurance people were able to agree bets and contracts

---

[31] 'Policy' from the Italian *polizza*, meaning 'written evidence of a transaction'.

through negotiation based on a subjective understanding of probability, this was still some distance removed from a rigorous mathematical approach.

By the sixteenth century, however, mathematics was thriving in Europe. The conditions were right for launching a mathematical attack on the subject of chance. Some centuries before, in 1202, Fibonacci had performed a great service to Western thinkers and mathematicians by introducing Hindu-Arabic numbers (our digits 0 to 9 and the place-value system), making serious arithmetic possible. With the Renaissance had come a flourishing, not only of arts and architecture, but also of mathematics. Copernicus had reframed the geometry of the solar system. Mercator had been regularizing map-making. Napier was inventing logarithms. Descartes was soon to turn geometry into algebra. Mathematical thinkers were revelling in the potential for applying the newly learned techniques of algebra to problems of all kinds, and they had a sense of the power that could come through symbolic and numerical treatment of these problems. It must have been intoxicating. Thinkers were filled with self-belief, and excited by the idea that mathematics could help to describe and understand the world in an unprecedented way, a way that often delivered demonstrable benefits.

Against that background, several minds started thinking about applying mathematics to the rolling of dice, and working out the essential rules of the probability game. The chronology poses a problem here, for the earliest thinking and writing that we know about on this matter was done by Gerolamo Cardano in the mid-sixteenth century, but his work was not published until over a hundred years later. So the work that is now recognized as the starting point of the intellectual development of probability theory is that of Blaise Pascal and Pierre de Fermat. In a context as seemingly frivolous and unthreatening as a puzzle arising from a game played at a table, these two mathematicians (in this context, their names are always linked) worked together to solve an intriguing intellectual challenge, and came away with the fundamental principles of modern probability theory.

So historians generally mark the birth of the theory of probability with the exchange of letters, starting in 1654, between Pascal and Fermat, in which they tackled a much-debated problem from the world of gambling, one posed to Pascal by gambler and writer, Antoine Gombaud. The Problem of Points, as it was known, arises when a game of chance (it matters not exactly what game) between two players is being played over a fixed number of rounds. Each has contributed money to a pot, and on completion, the pot will go to the winner, the first player to reach a targeted number of wins. Now suppose the game is interrupted. The question is: what happens to the stakes? How is the pot to be shared? You could simply divide the pot evenly, but natural fairness suggests that the player in the lead should get a larger share, but perhaps not everything, since their opponent still had a chance to win. How much, then? What would a fair division look like?

In their exchange of letters, Pascal and Fermat explored this and related questions from a number of viewpoints, working initially from specific cases but finally arriving at a generalized formula expressing the logic of their reasoning. They focused not so much on the rounds already won, but on how close each player was to the finishing line. For example, if the players (let's call them Left and Right) had won, respectively, three and four rounds each and the target was 5, then there would have been at most two rounds left to play. Let's see the different ways that those two rounds could play out. If we denote a win for Left by L and a win for Right by R, then we have these possibilities for those two missed rounds:

> L, R: which means an overall win for Right,
> L, L: an overall win for Left,
> R, R: an overall win for Right,
> R, L: an overall win for Right.

In the last two cases, of course, the final round need not be played at all; the match is already won for Right, but Pascal and Fermat correctly concluded that to ensure that all cases are treated equally, they needed to be listed and counted separately.[32] So in this example, three of the four outcomes lead to a win for Right, one to a win for Left; so a 3-to-1 split of the pot in Right's favour is appropriate. These are the odds of an overall win for Left, assuming each player has an equal chance of winning each round. If you were a bystander, making side bets on the game, you might conclude that Right had three paths to victory, Left only one, and you would agree that the odds were 3 to 1 in Right's favour. Betting on Left, you would expect to be offered 3 to 1 as fair odds.[33]

We can also reach the same result with the approach we used for working out the chances in craps. For Left to win, two separate, independent rounds must be won. Assuming a 50 per cent chance for each, the combined probability is

---

[32] This point, that their approach involved thinking about 'unnecessary' games, attracted some debate, but this is essential if you are considering all the equally likely paths from the given position.

[33] It's been speculated that the 'odds' terminology might arise from military battles. In the same way that one army might outnumber another 3 to 1, so the favourable outcomes for Right (in our example) outnumber the favourable outcomes for Left by 3 to 1.

25 per cent. For Right to win, there are two routes. Right could win the first round (with 50 per cent probability) or lose the first and win the second (this sequence has a 25 per cent probability). Overall, then, Right has a 75 per cent chance of winning and Left has a 25 per cent chance.[34] This approach, relying on the rules of probability theory, is consistent with the counting-of-ways approach that Pascal and Fermat used. This is unsurprising, since that theory arises precisely from the ideas of counting equi-probable outcomes.

While Fermat and Pascal's correspondence provided the first formulation of the modern approach to probability, it was not the first attempt to quantify chances mathematically. As mentioned, the polymath Gerolamo Cardano[35] (1501–1576) worked out some of the basic principles of probability, and set them out in his book *Liber de ludo aleae* (Book on Games of Chance). His work in this field did not achieve the recognition it might have, since this work was published posthumously in 1663, only becoming public some nine years after Pascal and Fermat had broken the ground. In particular, Cardano had identified and articulated the crucial principle of finding equally likely outcomes and counting them, to determine the proportion that were favourable. Cardano's book does not match our idea of an academic study: it is more like a gambling manual and includes rules for card and other games as well as games based on dice. It also contains a section on cheating, telling the reader about marked cards, and unfair dice, as well as a discussion of the rights and wrongs of gambling (arguing, for example, that playing at dice and cards could be beneficial in times of grief).

Shakespeare's nearly exact contemporary, Galileo Galilei (both were born in 1564, the likely year of the writing of Cardano's book), also figured out the basic rule for combining simple probabilities—multiply the probabilities of independent events. Both Galileo and Cardano had concerned themselves with 'the problem of three dice' that introduces this chapter. Just as we puzzled out the odds of various rolls of two dice in the previous chapter, they were chasing the logic behind the proper calculation of the probability of rolls of three dice. Both arrived correctly at the crux of the problem, that of counting the number of distinct ways that various totals can arise, just as we did earlier when considering the rolls in craps.

---

[34] And we are satisfied that they add up to 100%.

[35] Among Cardano's achievements: early adoption of negative numbers, invention of the combination lock, licence to practise medicine, cryptographic innovation, receipt of a life annuity from the Pope. He was also imprisoned by the Inquisition and, incidentally, his father was a friend of Leonardo da Vinci.

The Dutch scientist and mathematician Christiaan Huygens (born 1629) heard about the correspondence between Pascal and Fermat, but he did not have direct access to their letters. However, knowing of the discussion, he turned his attention to the problems they had been considering and, having reached the same principles, took the thinking a little further. Once again, his work was based upon games of chance, as reflected in the title: *The Value of All Chances in Games of Fortune* (1657). In his preface, he modestly acknowledges that 'for some time, the best mathematicians of France have occupied themselves with this kind of calculus so that no one should attribute to me the honour of the first invention'. Huygens addressed a broader range of questions, tackling the probability of winning in multi-round games, the determination of a fair price to pay for entry into a game, the division of the pot in an interrupted game, and an extension of this principle to three players. Although this is a broader range of problems, the context is still firmly the world of games and gambling.

Jacob Bernoulli, from a large family of mathematical thinkers, brought together the works of these pioneers in a comprehensive work called *Ars Conjectandi* (The Art of Conjecturing), published in 1713. Bernoulli clearly identified the utility of this work beyond the gaming table. He defines his subject as:

> The art of measuring, as precisely as possible, probabilities of things, with the goal that we would be able always to choose or follow in our judgments and actions that course, which will have been determined to be better, more satisfactory, safer or more advantageous.

In that sentence he captured the heart of the matter: that a sound understanding of chance can lead to better decisions and better outcomes.

Bernoulli's book consisted of four parts. In the first, he included and extended Huygens's work. The second part laid the basis for what we would now call combinatorics, the systematic mathematical counting of the ways that events can happen that is so essential to probability theory. Part three applies the principles to games of cards and dice. Finally, in part four, Bernoulli breaks free of the tidy world of games, to consider the application of the principles of probability to decisions beyond the realm of game-playing, in personal life, in judicial decision-making, and in finance.

The book also contains what Bernoulli called his 'Golden Theorem'. It's close to what we would now call the weak law of large numbers, which, broadly stated, says that with enough repetitions of a trial (a flip of a coin, a roll of a dice), the average outcome approaches the underlying theoretical probability, and that you can reason back from the results observed to reach conclusions about the underlying truth, the 'model' underlying the observations. Bernoulli's book, which was to prove hugely influential, also for the first time starts using the word *probabilitas*.

The name Bernoulli is associated with many ideas in maths and science. The Bernoullis were an extraordinary extended family, and collectively their achievements were pioneering in many scientific fields. But Jacob was the statistician. So, when, in the previous chapter, we considered Sarah's chances of emerging a winner after a series of bets on red, the mathematics used was that developed by Jacob. If Sarah had wanted to aggrandize her gambling habits, she could have referred to her bets as 'Bernoulli trials'—repeated random events, each with a binary outcome and constant probability.

For thousands of years people had actively engaged with chance, and yet it was not until the sixteenth century that a mathematical framework for probability was devised. It's intriguing to reflect on this breakthrough in thinking: how we move from Aristotle's 'unknowable' to the precise mathematical concepts that Galileo, Cardano, Pascal and Fermat, and Huygens figured out.

Why was the early development of the theory of probability so firmly grounded in game-playing and gambling? Perhaps we should not be so surprised. It could be that the gaming context was essential. Only in that artificially simplified world could the elegant fundamental principles of probability stand out from the many and varied confounding factors that muddy the waters in daily life.

For a start, a game (and the equipment for playing the game) offers the opportunity for repeated trials, an opportunity not presented by many other chance events. Secondly, the evident symmetry of the cubic dice (as opposed to, say, the asymmetric knucklebone) leads naturally to a consideration of six equal possibilities, and hence to the central principle of enumerating equally likely outcomes. Lastly, the clarity of the rules of the game, and the definite nature of the outcomes, means that there is no debate about the result, no uncertainty of measurement, and this means that attention can be focused on the random process itself.

Perhaps, too, social factors had their role to play. The gamblers who posed the questions that started the mathematicians on this hunt for the theory of chance must have spent many hours at the gaming table.[36] Gambling on this

---

[36] For example, to know from practical experience that throwing a 10 with three dice is more likely than a 9, is to be able to distinguish a probability of 12.5% from 11.6%. To gain that insight would have required many rolls.

grand scale would have required considerable amounts of wealth and leisure time. For gambling to be considered as a proper topic for polite academic discourse there needed to be a culture that tolerated vices of that kind. And to make calculations that would presume to predict the future would have required an intellectual irreverence that was not cowed by superstition or religious restraint.[37]

And as we know, probability is not just a game. The concepts, the relations, and the principles that underlie the mathematical analysis of games of chance are the same concepts, relations, and principles that underlie probabilistic approaches to subjects ranging from life insurance to political punditry. Whenever we think in a mathematical way about chance, we are relying upon concepts developed for and by gamblers.

Answer to:
Count the ways, and then count again

If there are six combinations that make a total of 9 with three dice, and also six combinations that make a total of 10, how is it that 10 comes up just a little more frequently than 9?

The answer, as Galileo could explain, lies in the number of ways that those combinations could be made. A dice roll with three different numbers showing can be made in six different ways. For example, $2 + 3 + 4$, making 9, can be rolled in these six ways:

(2, 3, 4), (2, 4, 3), (3, 2, 4), (3, 4, 2), (4, 2, 3), and (4, 3, 2).

On the other hand, a combination with two dice the same can be rolled in only three ways. $2 + 2 + 5$, also making 9, can be rolled as (2, 2, 5), (2, 5, 2), or (5, 2, 2). And any combination involving all three dice showing the same number can only be rolled in one way.

Looking at the ways in which a total of 9 can be rolled, we have:

$1 + 2 + 6$, (6 ways)
$1 + 3 + 5$, (6 ways)
$1 + 4 + 4$, (3 ways)
$2 + 2 + 5$, (3 ways)
$2 + 3 + 4$, (6 ways)
$3 + 3 + 3$. (1 way)

---

[37] Pascal himself was pulled between religion and mathematics, switching his point of view a number of times, but finally renouncing mathematics.

So, in total, 9 can be rolled in 25 ways.

What about combinations that make a total of 10?

  1 + 3 + 6, (6 ways)
  1 + 4 + 5, (6 ways)
  2 + 2 + 6, (3 ways)
  2 + 3 + 5, (6 ways)
  2 + 4 + 4, (3 ways)
  3 + 3 + 4. (3 ways)

And that makes a total of 27 ways to roll a 10, which is why it is slightly more likely than 9.

11 can be rolled in as many ways as 10 can, and those are the totals most likely to appear with three dice. Even so, the probabilities are relatively low: 27 out of 216 is a 1 in 8 chance: 12.5 per cent.

CHAPTER 3

Random Thoughts

I believe that we do not know anything for certain, but everything [only] probably.

Christiaan Huygens

Can you recognize randomness?

A coin is tossed 10 times. 'T' is noted if it comes up tails, 'H' if heads. I have two questions for you.

A) Which of these sequences is most likely to be the result of 10 coin tosses?
- T H H T H H T H H T
- T H T T H T H H H T
- H H H H H H H H H H
- H H H H H T T H H H

In fact, one of those sequences was indeed generated by my tossing a coin and simply recording the result. The other three were ones I invented. So:

B) Which do you think was the genuine random result?

Answers and discussion at the end of the chapter.

What are the Chances of That? How to think about uncertainty. Andrew C. A. Elliott,
Oxford University Press. © Andrew C. A. Elliott 2021. DOI: 10.1093/oso/9780198869023.003.0003

Randomness

random (adj.)

> 'having no definite aim or purpose'. From
>
> *at random* (1560s), 'at great speed' (thus, 'carelessly, haphazardly'). From
>
> > *randon* (Old French) 'rush, disorder, force, impetuosity', from *randir* 'to run fast'. From
> >
> > > ***rant** (Frankish root) 'a running'

Roulette etiquette allows the placing of bets up until the moment that the ball starts to slow down and begins to fall towards the centre of the wheel. After three rotations, the croupier declares 'no more bets'. But by that time, the ball is already on a course that is not entirely random. If you had sharp enough eyes, and quick enough wits, then based on the speed and position of the ball in the wheel at that last-but-one moment, you could place a bet that would have a better than 1 in 38 chance of winning.

That at least was the theory in the minds of Edward Thorp (he of the card-counting dispute with John Scarne) and Claude Shannon, a mathematician and the founder of information theory. They were convinced that once the ball's orbit in the wheel was established, the eventual destination of the ball was no longer a matter of pure chance, but would, with better-than-the-odds chances, fall into one of eight sectors of the wheel.[38] So in 1960, they ordered for themselves a standard Las Vegas wheel at a cost of $1,500 and installed it in Shannon's home. Then they set about developing the technology and the techniques to make their plan work. To do the predictions quickly enough to place a bet in the short time available would require them to smuggle electronics into the casino, and so they built a device, the size of a pack of cigarettes, activated by switches fitted into a shoe. This would allow the wearer to trigger a switch to record the timing of the ball as it made its first revolutions around the rim of the wheel. Accuracy to within 10 milliseconds was needed, but with training, they managed to get their reflexes fast enough to capture readings that were accurate enough. This timing information was then processed electronically, in order to predict one of the eight sectors into which the ball was more likely to fall. This prediction was transmitted via a wireless signal to another device in

---

[38] There's nothing magical about the eight sectors, it's just a convenient division of the wheel.

the pocket of the second conspirator. That player, most often Thorp, would then hear, through a concealed earpiece, a musical tone—one of eight possibilities—and would place the necessary bet.

Thorp and Shannon took their devices for a trial run to a casino in 1961. Their wives, Vivian and Betty, came along, partly to make it look like a normal social outing but importantly to act as look-outs. The mathematicians placed small bets only, and there were the teething issues that seem to accompany the use of any new technology. But when the device worked, it was successful at improving their odds. Thorp and Shannon did not break the bank, but they beat the odds. After the trial run, they called off the project for fear of the consequences, concerned about the possible reaction of angry casino operators. Nonetheless, they estimated that they had been able to shift the odds to 44 per cent in their favour.

In 1966 Thorp made this story public, and then in 1969 he was approached by a group of students wanting to replicate the project. It was not until 1976, though, that this successor project took off. The physics students from the University of California at Santa Cruz, who called themselves the Eudaemons, began work using considerably improved computer technology. The story, as recounted in the book *Eudaemonic Pie*[39] (Thomas Bass, 1991) is engrossing, as much for the account of the minimization of the technology (this time the computers employed were built into their shoes) and their encounters with casino security as for the analysis of probability and chance. As with Thorp and Shannon, the unreliability of the equipment and the inherent dangers of tangling with casino bosses meant there was never a 'breaking the bank at Monte Carlo' moment, but the principle was again proven: enough information can be gathered from early measurements of a roulette spin not only to overturn the house's edge but to create a significant advantage for the player. Don't think that this is a sure route to riches, though. In Nevada there has been, since 1985, a law in place prohibiting the use or possession of any device for betting prediction, a law directly attributable to the exploits of the wheel-clocking Eudaemons.

Where does randomness come from?

We see variation all around us. Some of it we can explain, some of it we can only find partial explanations for, and some we cannot explain at all. Randomness is what remains of variation when you have run out of explanations. It's the bit you can't get rid of.

[39] Published as *The Newtonian Casino* in the United Kingdom.

With a roulette wheel, we can understand where the randomness comes from. The speed of the wheel will be slightly different from one spin to another. The wheel's speed and the timing of the release of the small white ball are the input variables that Thorp & Co. sought to capture and whose effect they hoped to remove from the equation. Then there are tiny effects of friction that slow the ball and the wheel down, and finally, once the croupier has declared the betting closed, there is a chaotic sequence of bumps and bounces that finally directs the ball's path as its orbit decays and it spirals in to the centre. All these factors should be, on a fair wheel, unpredictable, uncontrollable, and therefore random enough, even if a computer in a shoe can go some way towards challenging this.

This is randomness arising from complexity. The path that the ball takes from the croupier's glove to the pocket where it comes to rest is complex, and therefore unpredictable. High precision measurement and accurate mathematical modelling can reduce the effect of that complexity and improve the odds, but some of the randomness will remain.

For some complex processes, the variables are simply too many, and the mathematics too hard, to expect very precise predictions. But there are some systems, some of them deceptively simple, where the variables are few and the mathematics is relatively straightforward, which are nonetheless unpredictable in practice. These are the systems described by chaos theory.

Chaos

chaos (n.)

> 'gaping void; empty, immeasurable space' (late 14c.,). From
>
> *chaos* (Old French, 14c.) or directly from Latin *chaos*. From
>
>> *khaos* (Greek) 'abyss, that which gapes wide open, that which is vast and empty'.
>
> Meaning 'utter confusion' (c.1600) from theological use of chaos in 'Genesis'. Chaos theory in the modern mathematical sense is attested from c.1977.

In chaotic systems, very small changes in the initial conditions result in disproportionately large changes in later states. Particularly where such systems involve feedback effects which amplify the differences, the state of the system becomes in practice unpredictable, even though the mathematics is fully understood. Chaotic systems are common. Those executive desk toys that use magnets to keep a pendulum moving in unpredictable ways? Chaotic. The way that syrup coils and folds when it's poured from a bottle? Chaotic. The chaos in these systems comes from the underlying mathematics: it is inherent, and cannot be engineered away. Short-term prediction may be possible, but in the long run these systems are always unpredictable. Even though they may be in principle deterministic, they can in practice be treated as effectively random.

Quantum

In one episode of the television drama *Chernobyl*, workers are tasked with opening up valves in the space below the unstable nuclear reactor's core, in order to drain water away and forestall the possibility of a massive steam explosion, should the structure give way and allow the water to become superheated. The soundtrack is a kind of static, a cascade of sharp clicks with no clear pattern. This is meant to represent the sound of Geiger counters measuring radioactivity.

A Geiger–Mueller counter detects ionizing radiation. A high voltage is set up within a tube filled with an inert gas. If radioactive materials are nearby, atoms of the gas will be ionized, and this will provide a path for an electrical discharge in the tube and cause a pulse of electricity. These electrical spikes are counted to measure the intensity of the radiation and hence the potential risk to health. The clicking sound comes from a speaker triggered by the count of ionization events. It sounds unpatterned because it comes from a source that is truly random.

The metal uranium is such a radioactive source: the atoms of Uranium-238[40] are unstable. That is to say, at any point in time there is a chance that an atom of uranium will emit an alpha particle made up of two protons and two neutrons. When this happens, the Uranium-238 atom will turn into a Thorium-234 atom. For a sample of material that contains Uranium-238, over a given period of time a known proportion of those atoms (on average) will decay in this way. A kilogram of Uranium-238 contains around 2.56 trillion trillion atoms. Every hour, approximately 45 billion of those (1 in every 56.5 trillion) will spontaneously undergo this change. For every hour that passes, every one of those uranium atoms has a minuscule chance, 1 in 56.5 trillion, of going 'pop' and decaying to a thorium atom. So over the course of 4.46 billion years,[41] each of those uranium atoms has a 50 per cent chance of decay. This period is known as the half-life of the isotope—the period it will take for half of the atoms to

---

[40] Uranium is the element, U-238 is the isotope. All uranium atoms have the same number of protons (92) but different isotopes have different numbers of neutrons. U-238 has 146 neutrons and is the most common isotope of uranium, but U-235 is what you need for a uranium-based atom bomb.

[41] Which happens, by chance, to be very close to the age of the Earth.

decay.[42] What tells a uranium atom that it's time to emit an alpha particle and decay? Nothing. It is random. A mere probability.

So this is pure randomness. Modern particle physics places probability and uncertainty at the heart of everything. When we look deep into the nature of the world, as deeply as we can, there we find probability. At the finest-grained level, nothing is certain. Particles have no definite location, but simply a probability of being in various locations. It's unsettling to know that, no matter how hard we tried, we could never eliminate uncertainty from our knowledge of the universe. But at the scale of human experience, due to the very large numbers involved, in most situations this quantum randomness is evened out and we experience a stable and predictable world. The randomness associated with the decay of atoms is a mathematically precise phenomenon, and the statistical properties that describe it are stable and measurable. While the decay of any one U-238 atom is entirely unpredictable, nonetheless when taken in very large numbers, the average rate of radiation can be calculated very precisely and will match observations very closely. Very large numbers can turn unpredictability at an **individual** level into near certainty at a **collective** level.

It's tempting to think that there might be as-yet-unknown mechanisms that, if we could just detect them, would provide an explanation for randomness at the quantum level. Famously, Einstein found it difficult to accept the idea of 'the Old One' playing dice with the universe: even that great mind felt that there must be an underlying cause, a reason, a model, a meaning. But mainstream modern thinking in physics is that there is no concealed clockwork that drives the timing of the decay of a uranium atom. Look as deep as you like: it's all probability down there.

Entropy

Do the shuffle

We've spun the roulette wheel and rolled the dice, but we've not yet shuffled the cards. The randomness of cards is not quite the same as the randomness of dice or spinning wheels. Cards dealt from a shuffled deck are random choices from a finite list of outcomes. If you deal the full deck, each possible outcome will have happened once and once only. One (and only one) of the four players sitting

---

[42] U-238 has a very long half-life. The more valued isotope, U-235, has a half-life of a mere 704 million years. Not all half-lives are as long. Carbon-14, used in archaeological dating, has a half-life of 5,730 years. Plutonium-238, used for uncrewed space missions, has a half-life of 88 years. Other isotopes have half-lives of fractions of seconds.

around the bridge table must be holding the ace of spades. So as cards are dealt from a shuffled deck, the probabilities change with each card dealt. That's why card-counting techniques in blackjack are effective: by tracking which cards have appeared, the card counter gathers some information about the distribution of the remaining cards. In a similar way, the shrewd bridge player will, over the course of bidding for and playing a hand, form an accurate idea of who holds which of the concealed cards. The randomness of a deck of cards is *revealed* when it is dealt and played, but that randomness is first generated when it is shuffled.

Entropy as a measure of uncertainty

Two couples, Alice and Bob and Carol and Dan, are playing bridge. Bridge is a card game played between pairs of partners who cooperate to win 'tricks'. At the start, the deck is shuffled and then cut to ensure that the cards are randomly distributed. The full deck is then dealt out to the four players in turn, each receiving a hand of 13 cards. Then, after negotiating a contract (more about that soon), the tricks are played. For every trick, each player plays a single card in turn, and the best card ('best' as defined by the rules of the game) wins the trick.

The number of possible orderings of a deck of 52 cards is very large: approximately 8.066×10^{67} permutations are possible.[43] Each time you shuffle the deck, you'll get one of these permutations, and if your shuffle is done thoroughly, the chance of any specific arrangement will be one in that very large number. It is overwhelmingly likely that a thorough shuffle will give you an arrangement of cards that is unique in the history of the world.

The physical quantity called entropy takes many guises, and one of them is in effect a measure of mixed-up-ness. In the case of our deck of cards, and knowing how many different arrangements there can be for that deck, we can put a number to the amount of entropy in the shuffled deck—how mixed-up it is (see 'Getting technical', p. 61, for details). This measurement is called Shannon entropy, named after the same Claude Shannon who went to challenge the roulette wheels with Ed Thorp, but whose outstanding achievement was the creation and development of information theory.

When a new deck of cards is broken out of its cellophane wrapper, the cards are in a specific order: there is no randomness. We can say that the entropy of this arrangement is zero. A single random cut of the deck will change this, but only by a small amount: the cards remain highly organized. Thoroughly shuffling the deck, though, will maximize the entropy.

---

[43] This is 52 factorial, written as 52!, and calculated as $52 \times 51 \times 50 \times 49 \times \ldots \times 3 \times 2 \times 1$.

Each player receives 13 cards in the deal. The act of dealing the cards does not change the entropy. But when the cards are picked up by the players, the usual practice is that each will then privately sort the cards they hold. The order in which their own cards were dealt to them is irrelevant (whether the king of diamonds was the first or the last card dealt to Alice matters not at all), so they sort them to make the hand easier to evaluate and to play. This sorting does not affect the game. There may be around 8×10^{67} possible arrangements of a deck of cards, but many of these will result in an identical distribution of cards into four bridge hands. In fact, there are 'only' around 5.36×10^{28} meaningfully different bridge deals.

By now, each of the players has some information: each knows exactly where one-quarter of the cards are, for they can see their own hand. For Alice there are now only 39 cards whose disposition among Bob, Carol, and Dan is unknown. From her perspective (and equivalently, from that of each of the other players), the overall uncertainty, the entropy, is less than for an outside observer who sees no cards.

From here on, as play progresses, uncertainty continues to be whittled away, but in a less precisely measurable way. In contract bridge, the game starts with the opposing teams bidding to be the ones to lead the play. The most ambitious bid secures a contract that obliges the winner to take a certain number of tricks or suffer penalties. In the bidding for the contract, information is shared between the players using a series of bidding conventions,[44] and in this way they can each deduce something more about the distribution of the cards. For example, Alice might deduce that her partner Bob is strong in clubs, that Dan has a weak hand, and that Carol has strength in spades. So for the players, the uncertainty about who holds which cards diminishes as the bidding progresses.

Perhaps Alice wins the contract (she is strong in diamonds). Her partner Bob now becomes 'dummy', and he lays his cards face-up on the table for Alice to use in play. This reveal of cards removes a further chunk of uncertainty from everyone's reckoning. Each player now knows the precise whereabouts of 26 cards, and has a vague idea of the distribution of the remainder. What Alice doesn't know is the precise split of cards between Carol and Dan, but based on the pattern of bidding she may have some clues about this, too. Part of the art of playing a bridge hand is sniffing out how the important cards are shared between your opponents.

Play begins and, as successive tricks are played, the cards held by Carol and Dan are revealed one by one. Many will be where Alice anticipates, but some surprises may still come her way. But it is sure that by the time 13 tricks have been played out, no mysteries will remain. It's a process of unravelling uncertainty.

---

[44] No explicit disclosure of the cards held is permitted.

Getting technical: calculating entropy

Claude Shannon formulated the theory that allows us to put numbers to the randomness of a shuffled deck of cards. His measurement, 'Shannon Entropy', is the log of the number of possible permutations of a closed system: it measures how many possible ways the elements of the system could be rearranged (*microstates*) and still be consistent with a large-scale description of the system (its *macrostate*). A deck fresh out of its wrapper is in one macrostate; a thoroughly shuffled deck is in another. The precise arrangements of the cards which are consistent with the macrostate description are the microstates. For the unshuffled deck, there is only one consistent microstate, while the thoroughly shuffled deck, where the cards may be in any order, may be in one of around 8×10^{67} microstates.

Shannon's entropy is the mathematical log of these numbers. For the unshuffled deck, it's the log of 1, which is 0. For the shuffled deck, it's the log (to base 2) of 8×10^{67}: 225.6. (It's conventional in information theory to use logarithms to base 2, and if we do this, the appropriate units of measurement are *bits*.)

When each bridge player sorts their own hand of 13 cards, this creates a new macrostate: a deck partitioned randomly into four hands in each of which the order in which they were dealt is irrelevant. How has the entropy changed? The number of different orders in which each player could receive 13 cards is around 6.227 billion. The sorting of the cards by each player is equivalent to reducing the overall entropy (of the overall arrangement of the cards) by 32.5 bits for each player. That leaves 95.5 bits of entropy remaining.

Each player, studying their own cards, gains further information. For each of them, that reduces the possible microstates by a factor of around 635 billion, equivalent to 39 bits. For each of them, there are now only 56.2 bits of entropy remaining.

When Bob, as dummy, lays down his cards, this reduces the remaining entropy to 23.2 bits. This is equivalent to around 10 million possible ways that 26 cards could be evenly split between two players. Then, as trick after trick is taken, the entropy reduces, bit by bit, until it reaches zero by the end.

Entropy as information

I've described that hand of bridge in terms of a progressive unravelling of randomness, and a corresponding reduction in entropy. It's a gradual reveal of the deal that resulted from one specific shuffle. There's another way of understanding that progression: in terms of information transfer.

The shuffled deck holds all the information that determines which player gets which cards. Every fourth card goes to each player, and when they pick up

their 13 cards, each receives a chunk of that information. Part of the appeal of
bridge lies in the negotiation of the contract, the cryptic communication
through which the players share information without ever disclosing any specific details. Then, when Bob, as dummy, lays down his hand, there is a further
information dump: everyone learns a little bit more. And as Alice, Carol, and
Dan play out their hands, the players finally gain the information they were
missing, until the cards hold no more secrets. Alice, had she been a professional
bridge player, would have had no difficulty in remembering every card that was
played, and would be able to recreate the four hands perfectly.[45] The information has been revealed.

Entropy in thermodynamics

But entropy did not start out as a measure of randomness or information. Rudolf
Clausius coined the term to apply it to thermodynamics: it represents the fact
that not all the energy in a system is available for mechanical work. If two sealed
chambers linked by a tube with a valve hold gas at different pressures, then
opening the valve will allow the gas to flow from one to the other, and that can
drive a turbine to perform useful work. Once the flow stops and pressures are
equal, no more work can be done. The equalization of the pressures has led to an
increase in entropy, and cannot be reversed except by adding more energy.

In Philip K. Dick's science fiction novel *Do Androids Dream of Electric
Sheep?*[46] we find this dialogue:

> [JR] Kipple is useless objects, like junk mail or match folders after you use the
> last match or gum wrappers or yesterday's homeopape. When nobody's around,
> kipple reproduces itself. For instance, if you go to bed leaving any kipple around
> your apartment, when you wake up the next morning there's twice as much of
> it. It always gets more and more.
>
> [Pris] I see.
>
> [JR] No one can win against kipple, except temporarily and maybe in one spot,
> like in my apartment I've sort of created a stasis between the pressure of kipple
> and nonkipple, for the time being. But eventually I'll die or go away, and then
> the kipple will again take over. It's a universal principle operating throughout
> the universe; the entire universe is moving toward a final state of total, absolute
> kippleization.

---

[45] But not, of course, the shuffled deck. She would have no information about the disordered
state in which each player found their cards.
[46] The book on which the film *Blade Runner* was based.

Dick's 'kipple' is a kind of entropy-made-physical. Useless, used-up objects, rather than useless, used-up energy. JR's comments about kipple apply to entropy, too. It requires energy to reverse entropy, and then only temporarily and in one place. And yes, the entire universe is moving to a final state where entropy wins and no energy remains available for useful work.

Entropy and the arrow of time

Much of the theory of physics is time-reversible. Watch a video clip of one pool ball striking another, then run it backwards: you'd not easily be able to tell which one was running forward and which in reverse. But zoom out and make a video of the first shot of the game where the cue ball breaks the triangle of stripes and solids, and you'll be in no doubt. Through time, order gives way to disorder. In large enough systems we can usually tell which way time is moving. The spilled milk and the shattered glass come after the rubber ball bounces onto the kitchen worktop, and not before.

What has this to do with chance? For a start, entropy is an explicit way of measuring uncertainty. It represents a mathematical way of measuring our ignorance of the state of a system, and that can be useful. The theory of information (Shannon's theory) that underlies all computer and most communications technology is a theory of uncertainty, and how communication transmits information that resolves uncertainty.

Then, too, entropy provides some explanation of why we perceive chance differently when we look forward in time, to when we look backwards. The ball bouncing across the kitchen was in fact on a course to hit the glass of milk, but before it happened, we didn't know for sure. The second law gives us the arrow of time, as one course is chosen from among all the possibilities available.

And not least, entropy is responsible for noise: variability without meaning. Most of the time, noise is an annoyance, and we take time and trouble (and energy) to eliminate it. But occasionally it can be useful. One of those uses is in generating random numbers.

Creating randomness

The fact that randomness requires a physical rather than a mathematical source is noted by almost everyone who writes on the subject, and yet the oddity of this situation is not much remarked.

Brian Hayes

Useful noise

In 1956 the UK Government introduced a new form of borrowing from the population: Premium Bonds. These were introduced by Harold Macmillan to encourage national saving, and were launched with the slogan 'Saving with a thrill'. The terms of these bonds included a guaranteed return of capital, but the 'thrill' was that the distribution of interest paid by the Government took (and still takes) the form of a lottery. Some investors would receive no dividends at all, others would be lucky and receive dividend payments of varying sizes. As might be expected, there was some opposition (Harold Wilson, leader of the opposition, called it a 'squalid raffle'), but the plan was an undoubted success, and continues to this day with a top prize of a million pounds, and over two million distributions of £25 prizes each month.

Something that contributed to the success of this populist innovation, and intrigued the people of Britain, was that the winning bonds were to be selected by an electronic random number generator, anthropomorphized as ERNIE (Electronic Random Number Indicator Equipment).[47] ERNIE was a hardware random number generator designed and built in the early days of computing. The machine was based on Colossus, the first programmable electronic digital computer, and was designed by Tommy Flowers and Harry Fensom, two of the creators of that machine.

ERNIE's random numbers came from 'noise' generated by neon tubes. All matter above the temperature of absolute zero is in constant motion at the molecular level. This restless jittering is effectively random: we analyse it, not by peering at the individual particles, but by looking at averages and variances. And the measurements we take show the effects of this chaos: the measurements are unstable, they wobble, and the effect of the wobble is noise. Noise is the 'snow' on an old-fashioned television screen. Noise is the hiss on a bad audio channel. Noise is the twinkle of a star as the wobbling atmosphere frustrates the astronomer trying to get a stable fix on it.

In the case of ERNIE, the source of noise was a set of neon tubes. In a neon tube, a high voltage is applied to electrodes at either end, and electrons start to flow between them. Along the way, the electrons may hit neon atoms, causing them to glow. As a side-effect, the path of the electrons becomes impossible to predict, and the result is a fluctuating current: a noisy signal. Usually noise is a

---

[47] ERNIE even made it into popular music. The lyrics to Jethro Tull's album *Thick as a Brick* include the line 'Good old Ernie: he coughed up a tenner on a premium bond win', and Madness have a song called E.R.N.I.E.

nuisance to be suppressed. But for ERNIE, the noise was what was wanted, and so it was trapped, amplified, and taken as a signal to drive circuitry that led eventually to the output of eight decimal digits and a letter, the format of the serial number of the bonds.[48]

Later versions of ERNIE used another natural source of entropy: thermal noise. ERNIE 4 made use of a chipset from Intel which generated random numbers from the thermal noise of transistors, using the heat generated from their operation. Most recently, in March 2019 ERNIE 5 was announced. It uses a quantum random number generator. At the heart of this is a stream of photons that strike a semi-transparent mirror. Which photons are reflected and which are not is a random behaviour in the purest quantum sense. The reflected signal can be captured and used as a source of random numbers, and ultimately determine which premium bonds get this month's dividend.

It's not just ERNIE that needs random numbers. Modern cryptography relies on randomly generated keys for secure communications. Hardware random number generators of the kind used in ERNIE 4 and 5 are now more generally available, and are used where elimination of security vulnerabilities is a priority. But randomness is not only useful in guarding secrets. It can also be critical to ensuring transparency.

Randomness beacons

A lottery draw is a public display of randomness. It's important that it be visible and recorded, since it might at some point be the subject of legal challenge. And lotteries are not the only situation where random numbers are used to make life-changing decisions. The United States green card programme works as a lottery, with a set number of visas being awarded to applicants in various categories, the lucky ones being chosen randomly. In these cases, too, it is important that the decisions be beyond legal challenge, and this can be ensured by using random numbers that are truly unpredictable but also publicly visible, and permanently recorded.

In 2018 the US National Institute for Science and Technology (NIST) launched its randomness beacon. This is a computer system that generates a new random

---

[48] In an episode of the law enforcement television drama NCIS, the plot centred on encryption, and the mechanism used to generate random numbers was based on the chaotic movements of fluids in a series of lava lamps. This is not simply a colourful dramatic invention: it is based on a real facility. The company Silicon Graphics did indeed develop a random number source based on images of a wall of lava lamps.

number every minute, and publishes it in a secure way. The randomness originates in a quantum source, an attenuated laser beam, which emits photons unpredictably. These photons are detected and the resulting signal is then processed in various ways to produce a 512-bit number, which is then recorded and published online, and so becomes a matter of public record.

If you're wary of using random numbers from that source, other options are available. The University of Chile publishes a randomness beacon based on a combination of independent sources including seismic measurements, local radio broadcasts, and Twitter posts. And if that's not enough, you can combine the outputs from different beacons, or even from your own source.

Pseudo-random

Anyone who attempts to generate random numbers by deterministic means is, of course, living in a state of sin.

John von Neumann

People go to a lot of trouble to generate truly random numbers—sequences that are impossible to predict. These are necessary wherever you plan to use randomness to guarantee secrecy or fairness. There are many situations, though, where what is needed are numbers that behave as if they were random but which don't really need to meet such strict criteria. In the classic computer game Tetris, a variety of shapes appears on the screen. It's important that they arrive in a 'random' order that cannot be anticipated by the player, but no state secrets are at risk. Similarly, in this book, I've relied on Monte Carlo simulations of random processes for illustrative examples. The fact that the random numbers behind the Tetris game or my examples are not produced from quantum effects is not important—what matters is that there is no obvious systematic pattern or bias to them that will spoil their purpose.

In principle, computer programs running in isolation should behave in completely predictable, controlled, and non-random ways. Nonetheless, randomness is essential for the proper operation of many computer algorithms. Games and puzzles need to present the player with unpredictable situations, but many statistical algorithms and modelling techniques (such as Monte Carlo methods) also require randomness.

In computer operations, it is often useful to distribute the load on different parts of the system in irregular ways and avoid synchronized requests for the same resource all arriving at the same time. For similar reasons, it may be advantageous to store data in ways that are not highly patterned. To design, test, and debug algorithms intended to cope with the complexities of the real world, you may need to simulate random inputs. In situations like these, computers can get away without needing an ERNIE-style hardware random number generator or the output from a randomness beacon but will instead rely on software to create what are known as pseudo-random numbers.

Before the availability of personal computing, and without sufficient resources to have your own ERNIE, a statistician might have relied on a published set of 'random numbers'. Here is an example:

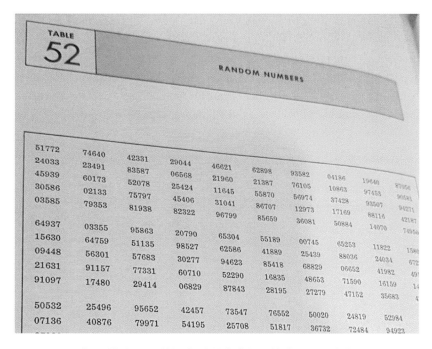

TABLE 52				RANDOM NUMBERS					
51772	74640	42331	29044	46621	62898	93582	04186	19640	87056
24033	23491	83587	06568	21960	21387	76105	10863	97453	90581
45939	60173	52078	25424	11645	55870	56974	37428	93507	94271
30586	02133	75797	45406	31041	86707	12973	17169	88116	42187
03585	79353	81938	82322	96799	85659	36081	50884	14070	74956
64937	03355	95863	20790	65304	55189	00745	65253	11822	1586
15630	64759	51135	98527	62586	41889	25439	88036	24034	672
09448	56301	57683	30277	94623	85418	68829	06652	41982	49
21631	91157	77331	60710	52290	16835	48653	71590	16159	14
91097	17480	29414	06829	87843	28195	27279	47152	35683	4
50532	25496	95652	42457	73547	76552	50020	24819	52984	
07136	40876	79971	54195	25708	51817	36732	72484	94923	

From *Mathematical Handbook*, M. R. Spiegel, McGraw-Hill, 1968.

But these days any computer language (and that includes spreadsheets) will allow you to access pre-written code that generates numbers that for most purposes are random enough. These pseudo-random numbers are entirely deterministic and completely predictable using known and repeatable methods, but they behave like random numbers. It's a nice paradox: sometimes absolutely deterministic data will serve as an acceptable substitute for randomness. Just as the successive digits of π look random but are no such thing, so it is with the random numbers that your spreadsheet's random function delivers.

Answer to:
Can you recognize randomness?

A) Which of these sequences is most likely to be the result of 10 coin tosses?
- T H H T H H T H H T
- T H T T H T H H H T
- H H H H H H H H H H
- H H H H H T T H H H

This is something of a trick question. Tossing a coin 10 times, there are 1,024 possible sequences that could result (2^{10}). Each of the sequences shown has precisely a 1 in 1,024 chance of appearing. We notice that the third option shown here has all heads, and that might lead us to think it would be unlikely to arise by chance. But every other option is just as particular as the run with all heads, and is just as unlikely to arise by chance. Each of them would occur by chance only 1 time in 1,024 on average.

B) Which one do you think was the genuine random result?

Again, something of an unfair question. In fact, it was the fourth option. I know—that run of five heads surprised me, too.

CHAPTER 4

Patterns of Probability

The statistician knows…that in nature there never was a normal distribution, there never was a straight line, yet with normal and linear assumptions, known to be false, he can often derive results which match, to a useful approximation, those found in the real world.

George Box

How long must he wait?

Jay is lucky: his place of work is close to the city art gallery, and as a member it costs him nothing to spend his lunch break there. Recently, though, he's been getting frustrated by one of the installations.

It's by Hamish Kipper, and it consists of a kind of cannon that fires a giant pellet of soft wax against the wall opposite on a random basis.[49] Here's how it works: a gallery employee sits by the cannon and every 10 minutes rolls a dice. If it's a 6, it's an indication to fire the cannon. The projectile splats against the opposite wall and oozes down to join an accumulation of soft wax. Everyone wants to see this happen, and so every 10 minutes a huddle gathers at the cannon to see if this time it will be fired. A chart is kept by the door to record each dice roll.

Now Jay has an hour for his lunch break, and he wants to share in the excitement. So every day, he heads straight for the Kipper exhibition and waits until he sees a 6 rolled, and the cannon fired. Jay's looked at the record of dice rolls, and he's worked out that the average time between firings is one hour, so he

[49] Apologies and thanks to Anish Kapoor whose wonderful piece *Shooting into the Corner* gave me the idea for this example. In the case of *Shooting into the Corner*, though, there is no random element: the cannon is fired on a fixed schedule.

What are the Chances of That? How to think about uncertainty. Andrew C. A. Elliott, Oxford University Press. © Andrew C. A. Elliott 2021. DOI: 10.1093/oso/9780198869023.003.0004

reasons that because he arrives at a random time between firings, he should expect to wait, on average, about 30 minutes. But it doesn't work out like that. He's starting to get into trouble at work for returning late from lunch. In fact, Jay's notes show that his average waiting time is almost an hour.

How can it be that he has to wait for an hour on average even though he arrives in between firings?

Answers and discussion at the end of this chapter.

Counting heads

Toss three coins at the same time and count how many heads you get. There are four possible outcomes—no heads, one head, two heads, or three heads—but not all of these four possibilities have the same chance of happening. By now, we know the drill: count the *ways* in which each possible outcome could happen. There is only one way in which the coins can land with no heads showing (TTT), but there are three ways in which one head could be showing (HTT, THT, TTH). There are three ways for two heads to show (HHT, HTH, THH), and there is only one way for them to land with all heads up (HHH). That accounts for all possibilities, eight in total.

So here are the chances of each possible throw:

Number of heads	0	1	2	3
Number of ways	1	3	3	1
Probability	12.5%	37.5%	37.5%	12.5%
Fair betting odds	7 to 1	5 to 3	5 to 3	7 to 1

Now consider this plan of a rather odd sequence of rooms:

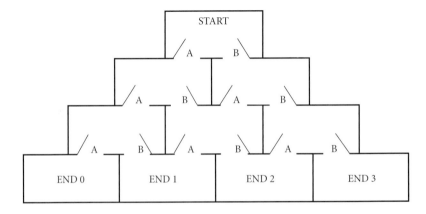

Imagine you're in the room marked START. The room has two doors marked A and B. Your instructions are to toss a coin, and if it comes up tails, choose door A. If heads, you should go through door B. Three times you face this choice, before you end up in one of the rooms marked END. What are the chances that you find yourself in each of those END rooms?

You'll have spotted that this is the coin-tossing game in a different guise. To end in room END 0, you must have thrown all tails. To end in room END 2, you must have thrown heads twice and tails once, and there are three routes you could have taken to get there: doors B, B, A; doors B, A, B; or doors A, B, B. It's the same distribution of probabilities as we saw with the coins: there is a 1 in 8 chance of reaching room END 0, the same for END 3. For rooms END 1 and END 2, the chances are each 3 in 8.

In the eighteenth century the pioneer of mathematical statistics Francis Galton (more about him later) designed a kind of toy to illustrate some principles of probability distribution. A stream of small balls is allowed to fall onto a diagonal array of pins. As drawn by Galton (on the left), there is only one pin that the balls falling from above can strike first: the middle pin of the second

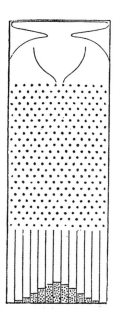
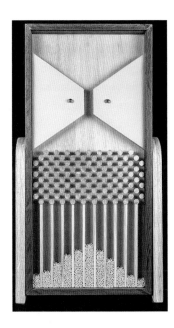

row. The ball will be deflected either left or right. At the next level, the ball having bounced one way or the other, there are two pins that could possibly be struck. The next level has three pins within reach, and so it goes on. The device, which carries the rather wonderful name 'quincunx',[50] is designed so that whenever the ball strikes a pin, it has an equal chance of bouncing to the left or the right. In fact, the first three bounces of a ball off the pins of the quincunx work in the same way as the choices you faced in the rooms in our strange building. In each case there is an equal chance of a left and a right choice. But Galton's pinball toy has many more rows, and many balls, and an arrangement at the bottom of collecting hoppers that can accumulate a stream of balls and so provide a kind of physical bar graph of how many balls end up in each end state.[51] Naturally, the precise distribution of the balls into the hoppers is different every time, but such is the statistical stability of the system that they tend to make a characteristic pattern, a gently curved mound. Relatively few end up at the outside edges; most accumulate in the middle. Which hopper a ball falls into corresponds to how many times it bounced to the right when it hit a pin on its downward journey.

Just as we can say of a single dice that each number has an equal chance of coming up, and we can say that to throw a 7 with two dice is three times as likely as throwing an 11, so too can we recognize that there is a stable pattern of probabilities associated with the balls bouncing down to the bottom row of the quincunx. If there was just a single pin, we would need only two hoppers, and on average they would each catch an equal number of balls.[52] If there were two rows of pins, we would need three hoppers, and they would catch balls in a ratio approximating 1:2:1. For three rows, four hoppers would be needed and the ratio would be 1:3:3:1. (This corresponds to our coin-tossing game.) Add another row, and we will need five hoppers, catching balls in the ratio 1:4:6:4:1. We can continue this pattern and arrange the numbers in a triangle. (We'll put

---

[50] Why this name? Quincunx comes from the Latin, and means five-twelfths. Roman coins having this value ($^5/_{12}$ of the coin called an *as*) were sometimes represented by five dots, laid out as four corners and a centre point—exactly the pattern used for the 5 on a dice. This arrangement of dots itself became known as the quincunx. Look carefully at the layout of Galton's pins and you will see that the pattern is based on a repeating quincunx-style arrangement. Oh, and the 'unx' part of quincunx comes from the Latin *uncia*, one-twelfth, from which we get both 'inch' and 'ounce'.

[51] If you study Galton's drawing, you will note that there are many more pins than are needed. Apart from the top corners, which can never be reached, this design as drawn will allow pins to bounce off the sides and fold the tails of the distribution inwards.

[52] If there were no pins, a single hopper would do, and would collect all the balls.

a '1' representing a single hopper at the apex of the triangle; this covers the case where the number of pins is zero.)

$$1$$
$$1 \; 1$$
$$1 \; 2 \; 1$$
$$1 \; 3 \; 3 \; 1$$
$$1 \; 4 \; 6 \; 4 \; 1$$
$$1 \; 5 \; 10 \; 10 \; 5 \; 1$$
$$1 \; 6 \; 15 \; 20 \; 15 \; 6 \; 1$$
$$1 \; 7 \; 21 \; 35 \; 35 \; 21 \; 7 \; 1$$
$$1 \; 8 \; 28 \; 56 \; 70 \; 56 \; 28 \; 8 \; 1$$

...

Apart from the initial '1' at the top, each number in this triangle is equal to the sum of the numbers above it, to the left and right. This arrangement of numbers is usually called Pascal's triangle,[53] and it holds many mathematical wonders, but for now let's just note the following:

- The number of entries in each row (the width of the triangle) corresponds to how many hoppers we would need for quincunxes of different depths.
- The numbers in each position of each row tell us how many different ways[54] there are for the corresponding hopper to be reached.
- The total of the numbers in each row is a power of 2, and is the total number of ways that a ball could travel to any place in that row.
- It follows, then, that each number in the triangle, when divided by the row total, is the probability of a ball reaching that point in the quincunx.

So the chance of a ball reaching the middle hopper of a pinboard with eight rows of pins is $^{70}/_{256}$, or around 27 per cent,[55] and the chance of reaching the rightmost one is $^{1}/_{256}$, or around 0.4 per cent.

---

[53] Yes, named for the same Pascal who corresponded with Fermat and gave us the first mathematical formulation for probability, although he was far from being the first to discover this remarkable arrangement of numbers.

[54] When talking about probability, we often talk of the number of 'ways' in which things can happen. I love that in this example the different ways are literally physical 'ways', actual routes that the balls could take.

[55] So a ball can end up at the central hopper in 70 ways, 35 of them with a final 'left' bounce, 35 with a final right bounce. This logic allows you to track the numbers of ways all the way back to the top of the triangle. Each number is the sum of the two above it.

This, then, is another pattern of randomness. Not the randomness of a single hit or miss, but one of counting how many hits are likely to happen over a number of repeated events: how many heads show up when multiple coins are tossed, or how many times a ball bounces left or right.

When a roulette wheel spins or a dice is rolled, each possibility has an equal chance, and that is called a 'uniform' distribution. This arrangement at the bottom of Galton's quincunx, though, is not uniform. This pattern is known as the binomial distribution.[56] We won't get too hung up on these names. As with so many words used in talking about probability and statistics, there is a technically sound reason for the word 'binomial', but it doesn't contribute much to an intuitive understanding. The point is that randomness comes in different patterns. If you're trying to make sense of chance, it helps to know what pattern of randomness you might be looking at. In most cases, the names are just labels.

Statisticians, as you might imagine, become very familiar with many of these standard distributions. These have been thoroughly analysed, and you can readily find relevant calculations and techniques for working with them in the textbooks and in computer software. This is not a textbook, so we're not going to do an exhaustive review, but we will take a look at a few of the more interesting ones.

Waiting for a six

Christiaan Huygens in the seventeenth century turned his serious mind to what might have seemed like trivial problems posed by the dice-rollers of his time. An example: players were asking whether they should bet on the proposition that a 6 would appear within four rolls of a single dice. They wanted to know whether the chances were better or worse than even. Of course, you might roll a 6 on your first roll, but you might also roll 10 times or more without a 6 showing. Indeed, there is no limit to how many throws might be needed.

---

[56] In fact, the binomial distribution does not need the coins (or whatever else generates the randomness) to be unbiased, as long as the bias is the same every time. For example, the count of the number of aces showing when a handful of five poker dice is thrown is also covered by the definition of the binomial distribution.

Which is more likely: that your first 6 will appear on the first roll or on the second roll? We know the chance of a 6 on the first roll—1 in 6. For a 6 to appear only on the second roll requires two things to happen: the first roll *must not* be a 6 (which happens ⅚ of the time), and then the next roll *must* be a 6 (⅙ chance). Taken together, this gives a chance of $^5/_{36}$, a smaller number than the chance of a 6 on the first roll. The chance of the first 6 appearing on roll three is again a little smaller, and for it to first show up on the fourth roll, we must first have failed three times, and only then succeed. As it turns out (see 'Getting technical', p. 76, for details), if we add together the chances of it showing up on any of the first to the fourth rolls, we end up at about 52 per cent, just a little better than even chances. So, if you're offered an even-money bet that a 6 would appear within four rolls, those would be favourable odds.

Here's a visualization of those chances in a simulation over 1,000 tries:

Red squares represent a 6 on the first roll, and there are 165 of them; green squares are second-roll 6s: 149 of them; blue squares are third-roll 6s: 116 of them; and the 110 purple squares are where a 6 appears for the first time on the fourth roll (grey squares are where no 6 appears within the first four rolls). There are 540 successes out of 1,000 in this simulation.

Once again, look at how surprisingly uneven the pattern is: we intuitively expect some kind of **uniformity**; instead, we get **variability**, with plenty of lucky and unlucky streaks. To get a clearer idea of the relative proportions of each possibility (but not the variability) we can rearrange those coloured squares into blocks. It's easy to see how first-roll 6s are more likely than second-roll ones, and how that pattern continues.

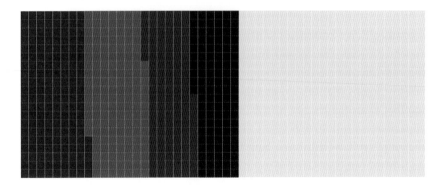

Getting technical: when does the first 6 show up?

For a 6 to appear on the nth roll, we must first have rolled something else $n - 1$ times, and only then succeed. Failing to roll a six has a 5 in 6 chance—so we can reason that the chance of a 6 showing for the first time on, say, the fourth roll is $\frac{5}{6} \times \frac{5}{6} \times \frac{5}{6} \times \frac{1}{6} = \frac{125}{1296} = 9.65$ per cent. Here are the chances of a 6 showing up for the first time in different positions in a sequence of rolls:

	1st roll	2nd roll	3rd roll	4th roll	5th roll	6th roll	nth roll
As fraction	$\frac{1}{6}$	$\frac{5}{36}$	$\frac{25}{216}$	$\frac{125}{1296}$	$\frac{625}{7776}$	$\frac{3125}{46\,656}$	$\frac{5}{6}^{(n-1)} \times \frac{1}{6}$
As percentage (shows up on that roll)	16.67%	13.89%	11.57%	9.65%	8.04%	6.70%	
Cumulative percentage (shows up then or before)	16.67%	30.56%	42.13%	51.77%	59.81%	66.51%	$1 - \frac{5}{6}^{n}$

Each fraction is $\frac{5}{6}$ of the previous one, which makes sense: the only difference between a 6 showing for the first time on the fifth roll rather than the fourth is that it must fail one extra time, and that extra failure has a probability of $\frac{5}{6}$.

The cumulative chances shown are the chances that a 6 has shown up by the time you get to that roll. So by the time the fourth roll has been made, the chances of a 6 having appeared are more than 50 per cent. Notice also that even after six rolls, the chances of having rolled a 6 are still less than $\frac{2}{3}$ (if only by the slimmest of margins). This shows that if something has a 1 in 6 chance of happening, it does *not* mean that in six attempts it will inevitably happen.

Finally, it is natural to ask about the average number of rolls before a 6 appears. The calculation of this involves an infinite series (after all, there is no limit to how many rolls could be required), but the answer boils down to a simple and intuitively satisfying answer: 6.

Why does any of this matter? Unless you're a seventeenth-century gambler, the dice-rolling is not the point. The point is the reasoning, and this reasoning underlies any question along the lines of 'how long will we wait before...' in regard to random, countable events. If floods of a certain severity occur with known (or assumed) probability in any year, then we can answer questions about how many years we can expect to pass without a flood of such severity.[57] In a time of climate change, when probabilities of flood are changing, we can use this kind of reasoning to assess whether changes in the weather are simply random variations within a stable pattern of probabilities, or whether they represent evidence that the underlying probabilities are themselves changing.

Queueing

How many checkouts should a supermarket provide? If you're in line at 11.30 on a Saturday morning, your answer might be a simple 'more than this!', but if you're designing the layout for a refit of the local superstore, you'll want a more sophisticated understanding of when your customers arrive. If you're building a web service, it helps to know the patterns of the times of day that your customers connect to your website, to ensure sufficient computing power to handle the possible peaks. If you're staffing an emergency department at a hospital, knowing the patterns of arrival of patients will help you to schedule your staff rotas to best cover the load.

It's not enough to know the averages. If you could depend on your customers to arrive at a constant, steady rate (**uniformity**), then you could provide just enough to service the queues at the average rate. But the spikes in demand will catch you out, unexpected rushes of activity, and you'd better know how bad those spikes could get, and how often they could happen (**variability**). It's not enough to say 'it's unpredictable'. Uncertainty can be measured and managed.

The randomness behind these 'customer arrival' problems is often well modelled by a so-called Poisson distribution (after Siméon Denis Poisson, a nineteenth-century mathematician and the first person to describe it). The

---

[57] My father-in-law, a civil engineer, used to talk about designing a bridge to survive 'the fiftyyear-flood'.

Poisson distribution arises when you're counting how many events of a given type happen in a given time, if each event has a certain probability of happening in that time.

Think about the number of customers who arrive at the fishmonger's counter at your local supermarket between 10.00 and 11.00 on a typical weekday morning. Perhaps there's an average of 12 customers in that hour. Some days there will be fewer, and some days there will be more. The actual number is unpredictable, of course: it's a random number, but it's not from a uniform distribution. What are the chances that you'll have six or fewer customers in that hour? A calculation based on the Poisson distribution suggests around 4.6 per cent, which works out to about 1 time in every 22:

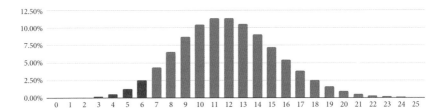

What are the chances of more than 12 customers? Around 42 per cent:

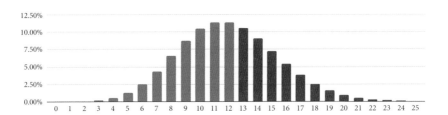

Continuous distributions

Not all statistical measurements have outcomes that can be written as whole numbers. Dice rolls, roulette spins, Geiger counter readings and numbers of customers buying fish can all be expressed as integers. But many of the statistics that we are interested in are measurements of continuous quantities, such as distance, or time, or speed, or mass. And these continuous quantities are also subject to random variation, and the laws of probability apply to them, too. When it comes to continuous quantities, we must think not about single numbers, but instead about ranges.

A traffic officer stationed on the side of the road uses a radar gun to measure the speeds of cars approaching them along the highway. What is the chance that a car will be travelling at a speed of precisely 120 km/h? Virtually nil, and the more precise their radar gun, the more decimal places it displays, the smaller the chance that it will display exactly 120. We need instead to ask about the chance that the reading from the device is greater than 120 km/h.

For continuous random quantities, too, there are some patterns of probability that come up over and over again. And one stands out above all others: the 'normal' distribution.

Pascal's triangle taken to the limit? That's normal

Here's the 12th row of Pascal's triangle:

1 12 66 220 495 792 924 792 495 220 66 12 1

Recall that those numbers represent how many ways a ball can reach each of the hoppers at the bottom of a quincunx. Divide those numbers by their total, 4,096, and we have the probability that a ball falls into each of those hoppers:

0.02% 0.3% 1.6% 5.4% 12.1% 19.3% 22.6% 19.3% 12.1% 5.4% 1.6% 0.3% 0.02%

If we show those numbers as a line graph, it looks like this:

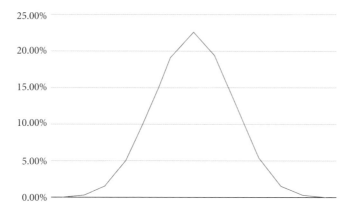

As Pascal's triangle grows, the shape described by the numbers in the lowest row becomes more and more like a smooth curve, curving gracefully upwards, over the top of a rounded peak and then mirroring the ascent with a matching downward arc. We can still interpret those numbers as probabilities, as the chances of balls ending up in smaller and smaller hoppers, after a larger and larger number of pinball bounces, left and right. What if we take this to the limit? Can we think about the chance of ending up in an infinitely narrow hopper, after an infinite number of vanishingly small bounces one way or another? We can indeed, and, taken to the limit, this leads to the most famous of all continuous probability distributions, the rock star of statistics. It's the one known variously as the Normal Distribution, the Gaussian Distribution, or the Bell Curve.[58]

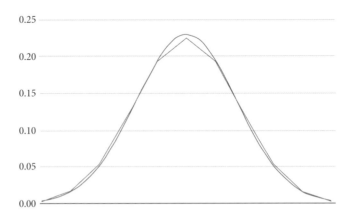

Comparing the binomial (blue: a quincunx with 12 layers) to the normal (red).

Everybody believes in [the Normal distribution]: the experimenters, because they think it can be proved by mathematics; and the mathematicians, because they believe it has been established by observation.

Henri Poincaré, *Calcul des probabilités*

There are two related reasons why this distribution is so famous:

The first is that the normal distribution crops up so frequently. The distribution of many (but not all) measurements in practical work can be well approximated by this standard distribution. The heights of adults, the lengths of normal human pregnancies, the size of errors made when taking linear measurements[59]—all are pretty well approximated by the normal distribution. The

---

[58] Statistical pioneer Francis Galton was the first to use the term 'normal' in the sense of conforming to this standard distribution. Karl Pearson, his pupil, claimed to have coined the term: he was wrong, but Pearson's consistent use of the term led to its widespread adoption.

[59] A mathematical function closely related to the normal distribution is in fact called the error function.

starting point for a statistical analysis is often to think about whether or not you are dealing with something that is close enough to being a normal distribution. If so, you're in luck: you'll have plenty of tools at your disposal for dealing with it.

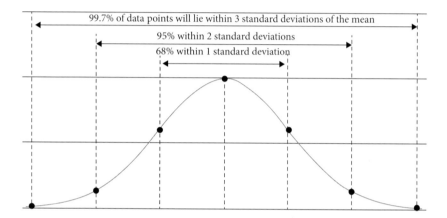

99.7% of data points will lie within 3 standard deviations of the mean

95% within 2 standard deviations

68% within 1 standard deviation

We talk of *the* normal distribution, but really it's a family. We choose which member of the family we want by specifying where the centre lies (the *mean*), and by how spread out the distribution is (the *standard deviation*). Once we've done that, we've nailed it down. Then for every member of the family, all of the following are true. Take a random data point from that distribution:

- There are equal chances of it being greater than or less than the mean: 50 per cent.[60]
- There is about a 68.3 per cent chance that it is less than one standard deviation away from the centre. Conversely, there is about a 31.7 per cent chance that the number is further from the centre (the mean) than one standard deviation.
- There is about a 95.4 per cent chance that it is less than two standard deviations from the mean, and only about a 4.6 per cent chance that it is further than this.
- There is about a 99.73 per cent chance that it is closer than three standard deviations from the mean, and only about a 0.27 per cent chance that it is further than this.

---

[60] The chance of being equal to the mean? Zero, to as many decimal places as you like. When we are in the continuous world, we need to talk in terms of ranges.

Maybe we can visualize that in a different way: here's a simulation of 2,000 data points from a normal distribution, coloured according to how many standard deviations away from the mean they are:

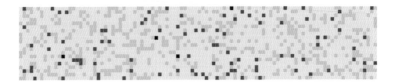

The very dark cells (there are only four of them) are more than three standard deviations away from the average (the distribution suggests there should be around five like this); the dark red ones are outside two standard deviations, and the pink ones are outside one standard deviation (less than a third of the cells). Around ⅔ of the cells are grey: these are within one standard deviation. This shows how quickly the normal distribution tails off, how few cells have extreme values. If you've assumed that your data comes from a particular normal distribution and you find a value that is more than three standard deviations from the centre, that's like coming across one of those black cells. They're rare and you should expect only around 3 in 1,000 to be that far from the central value. Your assumptions might be worth a second look.

The second reason that the normal distribution is so important is that it has a quite remarkable property which in part explains why it is so ubiquitous. Here's the thing: if you add together or average together sufficiently many measurements from other (not necessarily normal) distributions, the total (or average) of those measurements will start to look like a normal distribution, and the tools used to work with the normal distribution will do a reasonable job on those totals, or averages. This is astonishing, and I'll try to explain why.

Let's take an absurd example. I have on the shelf near my desk a set of dice used for playing the game of *Dungeons and Dragons*. The rules of that game often involve invoking chance using probabilities other than the 1 in 6 provided by a standard dice. My set of dice contains one 20-sided dice, one with 12 sides, one with 10, one with 8, one with 6, and one with 4. Further, as it happens, I have three coins on my desk. Let's suppose that I gather them all together and cast them on the desk in front of me. Then I count up the total, counting dots for the dice and heads for the coins. It's a quite nonsensical mixture of randomizers, enough to make a purist shudder. The maximum total I could get from those dice and coins is 63, the minimum is 6 and the average is 34½. Now, I'm not going to throw and count these 1,000 times, but I can do the Monte Carlo trick and simulate 1,000 throws in a spreadsheet. So that's what I did. Here's how my results compared to the predictions of a scaled normal distribution:

Total count	Normal distribution predicts	I got
Within 1 standard deviation of 34.5	683 out of 1,000	665
Within 2 standard deviations	954 out of 1,000	946
Within 3 standard deviations	997 out of 1,000	999
Greater than 34.5	500 out of 1,000	497

In other words, by combining random numbers from distributions that really have no business being added together, where some elements outweigh others, I end up with a statistic (the total of all my dice and coins) that actually fits the normal distribution quite well.

Getting technical: Central Limit Theorem

This remarkable and powerful feature of the normal distribution means that, even when we do not have any detailed knowledge of how some measured quantity is derived or what internal components go towards making it up, we can get useful and quite reliable answers by working with those measurements that we can observe, as long as we have reason to believe that the component parts are combined by adding up. This property, that when you add together random variables the resulting distribution starts to look like the normal distribution, carries the impressive title of the Central Limit Theorem. It's one of the ways in which order and predictability in a big picture view can emerge from randomness and uncertainty at a more detailed level. It opens the door for us to gain clarity from chaos.

The normal distribution is extremely useful. But when you have a hammer, everything looks like a nail. The danger of discovering a highly useful tool is that you might overuse it. The normal distribution is in many circumstances a safe enough assumption to make, and especially in the middle of the distribution, where most observations fall, it is often a good fit for observations. Out at the extremes, though, in the tails of the distribution, it can be much less reliable.

In particular, in the financial world a body of statistical work was developed around the pricing of some securities that assumed a normal distribution for price movements in financial markets. It turns out that this is *not* a very safe assumption. In particular, it underrepresents the extremes of the data: prices of investments move by large amounts more often than the normal distribution would predict.

But many random variables do fit the normal distribution closely enough to make it a valuable tool. To characterize a random variable using the normal distribution, you simply need to know where the peak is located (the mean) and how wide the spread is (the standard deviation). If you have a fit, then you can expect about two-thirds of cases to fall within one standard deviation of the mean, and almost all cases to fall within three standard deviations. Those rules of thumb go a long way to helping you to answer the question 'what are the chances of that?'

The ideal and the approximate

All models are wrong, but some are useful.

George Box

The mathematics of the casino are clear-cut. The models of probability that we use to work out the odds in games of chance are very good matches to reality. But in the more complex, uncontrolled world outside the casino, we have much less reason to be confident that we understand all the factors that make life random.

All the distributions we've talked about in this chapter are ideal models that fit reality imperfectly, to a greater or lesser degree. Even Galton's pinball game will not conform exactly to our assumptions: the pins are unlikely to be placed with absolute precision; a bump to the table while the balls are falling will disrupt the ways in which they trickle into the hoppers; sometimes the ball bounces up and goes on to hit another pin at the same level. Very few real situations will perfectly meet the theoretical criteria for using our statistical tools. Nonetheless, these patterns of probability prove themselves useful, time and again, and allow us to make calculations about the random things in our world that turn out to be surprisingly accurate, and equip us to engage with uncertainty in an effective way.

Answer to:
How long must he wait?

This is one of those near-paradoxes of probability that remain intriguingly baffling even after you've managed to understand the workings.

The key to it is realizing that the dice rolls that the museum attendant makes are independent of each other, and that means that the chances of a 6 appearing remain the same, regardless of how long a gap there has been since the last one.

The rolls that occurred before Jay arrives, and their timing, are entirely irrelevant to the waiting time. It's just as if the installation opened each day at the moment of Jay's arrival. Every time, just before the attendant makes the roll, the expected time before the next firing of the cannon is the same. It makes no difference whether the previous successful roll was recent, or if there has been a long wait beforehand.

For the same reason, a slot machine that operates on a purely random mechanism will offer the same odds of winning, regardless of how long it has been since it last paid out. Random processes like this have no memory: the false intuition that they might do is a version of the Gambler's Fallacy.

The second duality: Randomness–Meaning

I'm playing a board game with my family and my cardboard forces are mounting an attack on those of my son. To resolve the conflict, I must roll a dice. I know that the outcome will be random. And yet I feel it shouldn't be. I've lost similar rolls twice in a row. Aren't I due for some luck? I want this to be a story of setbacks overcome and victory gained. I want this roll to mean something.

Much that happens around us is predictable and understandable. I drop a rock: it falls. Cause and effect. The rain doesn't come and the crops fail. It seems that the rain causes the crops to grow. But what causes the rains to fail? There is no easy explanation for good rains one year, and drought the next. Rejecting mere chance, we crave explanation. Perhaps the gods are angry.

We love stories, and from the earliest myths to last night's crime drama on television, the tales that please us are those of coherent events driven by chains of cause and consequence. Random happenings make for bad stories, and so we create myths and superstitions to fill the gaps where we have no other explanation. And can we really believe, some ask, that a virus like Covid-19 can appear randomly? Surely not: it must have been created in a Chinese laboratory and have been activated by radiation from 5G transmission masts, they declare.

Uncertainty can be pure chance, *aleatoric* uncertainty, where no good explanation is available, or *epistemic* uncertainty, where an explanation exists, but we just don't know what it is. We tend to prefer the epistemic kind of uncertainty: at least there is a reason, if only we could find it. Nassim Nicholas Taleb writes of the narrative fallacy, a tendency to concoct a story to explain a sequence of events which may not in fact be connected in any way. A story with its glue of cause and consequence is easier to remember than a jumble of unconnected events. 'Because' and 'therefore' are powerful words.

But there is a better way to extract meaning from the seeming chaos around us. Where our uncertainty is epistemic, stemming from ignorance and not from mere chance, we can address that ignorance by acquiring knowledge. Where the uncertainty is aleatoric, pure chance, we can understand that variability through statistics. These scientific explanations summarize understanding in the form of models. Good models give better results, and so good explanations can drive out the bad and remove the need for superstition and ritual, while always allowing the possibility of improvement.

PART 2

Life Chances

John Graunt: Watching the hatching and dispatching

For no man can be said to die properly of Age, who is much less [than age 70]: it follows from hence, that if in any other Country more then seven of the 100 live beyond 70, such Country is to be esteemed more healthfull then this of our City.

John Graunt

In 1676 the fifth, and final, edition of a volume entitled *Natural and Political Observations Made upon the Bills of Mortality* was published. Written by John Graunt (1620–1674), it was an analysis of the records, published weekly, of christenings and burials in London, and the first edition had been published in 1662. (The final edition was published only after his death, probably compiled by Sir William Petty.) In the preface to the book, Graunt notes the great public interest in these weekly notifications, but that for most people, their interest is drawn only to:

how the Burials increase, or decrease; And, among the Casualties, what had happened rare, and extraordinary in the week currant; so as they might take the same as a text to talk upon, in the next Company.

Graunt stumbled into his intense interest in the Bills ('I know not by what accident', he writes) and goes on to describe how he reduced many volumes of statistics into 'a few perspicuous tables', in order to undertake a more serious study of the data set he had discovered:

[W]hen I had reduced [the bills] into Tables so as to have a view of the whole together, in order to the more ready comparing of one Year, Season, Parish, or other Division of the City, with another, in respect of all the Burials, and Christnings, and of all the Diseases, and Casualties happening in each of them respectively; I did then begin, not onely to examine the Conceits, Opinions, and Conjectures, which upon view of a few scattered Bills I had taken up; but did also admit new ones, as I found reason, and occasion from my Tables.

In other words, he collated the data, organized it, and cleaned it. He then noticed patterns in the data, formed hypotheses, and set about examining them. He bemoaned inconsistent data recording (particularly in regard to cause of death), to which he strived to bring some regularity. The aims of his work, if not the manual methods, would be entirely familiar to modern data scientists. He then proceeded to draw conclusions. For example:

> That the wasting of Males by Wars, and Colonies do not prejudice the due pro-portion between them and Females: That the Opinions of Plagues accompany-ing the Entrance of Kings is false, and seditious: That London, the Metropolis of England, is perhaps a Head too big for the Body, and possibly too strong [...] That the old Streets are unfit for the present frequencie of Coaches.

Graunt was not a professional mathematician or statistician. (The term 'statis-tics' would not be coined until a hundred years after his death.) He was a hab-erdasher by trade, had started his working life in the family business, and had made such a success in commerce that he ended up holding prestigious offices in the Drapers' Guild. He became a connoisseur of art: Samuel Pepys, who knew him well, expressed extravagant admiration for his collection of prints. His pioneering statistical work appears to have been driven by curiosity and care for the good management of his city.

John Graunt, in dedicating his work to Robert Moray, President of the Royal Society, was modest regarding his academic credentials. But by virtue of the studies he made of the statistical records of London, he is regarded as one of the first demographers. After the publication of his *Observations*, he himself was admitted as a member of the Royal Society. But in 1666, his home and his busi-ness were lost in the Great Fire of London, after which he fell into financial difficulties, leading to bankruptcy.

Graunt died relatively young, at the age of 53. The cause of death recorded was jaundice and liver disease.

Life Chances

No one can avoid gambling, because life itself forces us to make bets on Dame Fortune. In business, education, marriage, investment, insurance, travel, in all the affairs of life we must make decisions which are gambles because risk is involved.

John Scarne

As we step away from the air-conditioned cocoon of the casino into the smelly street, we become aware of the barely contained chaos of the world. The wind blows, traffic blares, pedestrians running late for the next appointment jostle one another. Our days seldom go precisely to plan. No less than in the casino, here on the street it is a world where uncertainty rules. But this is not the uncertainty of the turn of the card, or the spin of the roulette wheel. Here it is the wayward turn of the steering wheel that can make the difference. This unpredictability arises not from crisp theories of probability, but from the uncontrolled complexity of life in all its glory and ugliness.

How are we to measure and manage the risks of life? Every moment of every day we are exposed to the operation of chance. You may slip on a banana skin. You may find a sack of gold coins under your floorboards. Even while sleeping, your DNA may suffer damage from carcinogens in the meal you ate, or in the cigarette you smoked. How can we manage the effects of chance upon our lives without first understanding our exposure to risk?

Exposed to risk

A friend asks you to guess the most dangerous sport in the world. A range of possibilities comes to mind: motor racing? base jumping? free diving! You're sure you've heard that free diving is the most dangerous sport there is. The answer, you're informed in a smug tone, is in fact lawn bowls, because more people die playing that sport than any other. She doesn't say it, but you can see the 'Gotcha!' in her eyes. You get it. It's a popular sport, and those playing tend to be older people. But the question annoys you. It makes you feel mildly foolish, and you're not quite sure that the word 'dangerous' should be taken to mean the sport that has the highest absolute death rate among participants. Shouldn't you at least allow for the numbers of participants? And recognize how regularly they compete? Now you're thinking along the right lines. To assess the risk fairly you should compare the number of deaths to the number of opportunities for someone to die while playing bowls. You need to work out how many times someone was exposed to that risk.[61] To measure risk fairly, we need to set the number of times an event happens against the number of chances there are for it to happen. One hospital performs 10 times more triple-bypass operations than another. It also has many more cases where complications have arisen. Are you surprised? Of course not. There were more opportunities for things to go wrong. Ten times as many patients were exposed to risk. We need to think in terms of rates and ratios.

---

[61] And maybe, too, take into consideration matters such as the age of those who play.

Proportion or probability?

If, in a class of 40 school leavers, 24 go on to higher education, that's 60 per cent of them. That's a statement of proportion. It has, in itself, no connection to probability; no uncertainty is involved. But it's very common to hear such a statistic expressed in terms of probability. Someone might say that there was a 60 per cent chance that those students would go on to study further.

But for any individual student that statement is meaningless. Jana always planned to go to university and the only uncertainty in her mind was where (and not whether) she would be accepted. Her chance of studying further was always much greater than 60 per cent. For entrepreneurial Stephan, though, who's intending to start his own business, there was always a zero chance of continuing his formal education.

We do it all the time, but turning a proportion into a probability conceals a hidden step of logic. It implies that *if we chose a member of the class at random*, there would be a 60 per cent chance that the one we chose would study further. It's a shift of perspective. By talking about the percentage, we show that we are not interested in Jana's or Stephan's outcome as an individual, but only in the class as a whole—perhaps in the form of an anonymous, representative, random class member. These two perspectives illustrate one of the dualities: **individual–collective**. This is a relatively benign example, but it's not so hard to see that similar logic can fuel prejudice. If supporters of the Red team are, statistically, more often convicted of violent offences than fans of the Blues, then it is easy to slip into saying that a Reds supporter is more likely to be a troublemaker. How much weight (if any) should this reasoning from group proportions carry when we make a judgement about an individual? We'll explore this in more detail later in the book, but for now it's enough to recognize that proportions and probability, though related, are not the same thing.

When we summarize the statistics in this way, we choose to throw away much of the specific information we have gathered, but we do so in the hope that the summarized statistics will be useful beyond the specific group we have measured. In a drug trial, the researchers might discover that 12 per cent of the subjects experienced an adverse side-effect from the drug. What do we conclude from this? That when we prescribe the drug in the future, it's reasonable to expect a 12 per cent chance of that side-effect occurring.

CHAPTER 5

Isn't That a Coincidence?

A report of holding 13 of a suit at Bridge used to be an annual event. The chance of this in a given deal is 2.4×10^{-9}; if we suppose that 2×10^6 people in England each play an average of 30 hands a week the probability is of the right order.

…This is of course a subject made for bores.

J. E. Littlewood, *Mathematical Miscellany*

Signs and portents

The author Mark Twain was born on 30 November 1835, at a time when Halley's comet was prominent in the sky. The comet has a period of around 75½ years and the recorded date of birth of Samuel Longhorn Clemens (Twain's real name) was just two weeks after perihelion, the moment of the comet's closest approach to the Sun. In his autobiography, published in 1909, Mark Twain wrote:

> I came in with Halley's comet in 1835. It is coming again next year, and I expect to go out with it. It will be the greatest disappointment of my life if I don't go out with Halley's comet. The Almighty has said, no doubt: 'Now here are these two unaccountable freaks; they came in together, they must go out together.'

The date of the next perihelion of the comet was 20 April 1910. The day after that, Twain died.

Coincidence?

Comments at the end of this chapter.

What are the Chances of That? How to think about uncertainty. Andrew C. A. Elliott,
Oxford University Press. © Andrew C. A. Elliott 2021. DOI: 10.1093/oso/9780198869023.003.0005

What's in a name?

A few years ago I boarded a flight to Cape Town, only to find another passenger in my allocated seat. There were free spaces for the other three members of my family, but my seat was occupied. Some mistake, surely? I compared boarding passes with the man sitting in *my* seat. Look, I said, you've got my boarding pass—it clearly says 'Andrew Elliott'. Of course it does, he replied, that's my name! What were the chances? The mix-up was quickly resolved and the flight passed without further incident. Now, as coincidences go, that was less than jaw-dropping. Still, it surprised me at the time, and I bored friends and family with that story for months. Should I have been surprised? What *were* the chances?

What are the chances?

Over 200 flights, with 400 passengers, and a probability of a name-clash with each of them of 1 in 30,000, on average there will be 2.7 name clashes.

Representative example:
80,000 chances for clashes
Observed clashes: 3 = 0.004% (expected average: 2.7 = 0.003%)

(If you look very closely, there are three dark pinpricks among those 80,000 cells.)

Nothing highlights the tension between personal experience of randomness and cold analysis of statistics quite as strikingly as a dramatic coincidence. It's another example of the **individual–collective** duality. And because coincidences arise from the swirl of events of daily life, they are difficult to analyse in a rigorous way. But let's make a stab at calculating the chances that I would have a name mix-up like that on a long-haul flight.

My name is far from being unusual. I have several times worked in organizations with others sharing my first and last names. A quick directory look-up gives me 35 hits on 'Andrew Elliott' in Surrey, my home county, which has around a million inhabitants. That's around 1 in 30,000 people who share my name.[62] If the travellers on a Boeing 747 flight from London to Cape Town were of a similar mix to the people of Surrey, then there would be around a 400 in 30,000 chance that one of them was called 'Andrew Elliott'. That's a 1 in 75 chance.

I don't fly very often, but there may have been around 200 occasions (say, eight flights a year for 25 years) when a similar mix-up could have occurred. 200 opportunities for a 1 in 75 chance? It starts to look like it was inevitable!

Coincidences

No doubt every one of us has a tale to tell of a weird alignment of dates, places, people, and objects that seems improbable to the point of being impossible. My sister, on 16 May 2019, consulted a recipe book inherited from my mother and found the sole annotation in the book, in my mother's handwriting. Alongside a recipe for Peppered Sirloin with Garlic Mayonnaise was written my sister's name, the date 16/5/91, and the word 'Good'. The coincidence has two parts: that it was my sister who found her own name, and that she found it on the same day of the year that it had been written. What could explain that?

I am sure that, whatever arguments I make that cases like this one are simply statistical outliers, and that we should expect our lives to contain a scattering of them, I will not convince all my readers, and probably not even my sister. The connection between my mother writing that note and my sister reading it exactly 28 years later is personal. It feels like it should be meaningful, even if the

---

[62] If I moved to County Fermanagh in Northern Ireland, I would be one of six Andrew Elliotts in a population of 62,000 (1 in 10,000). There are 270 Andrew Elliotts reported in the United Kingdom, a population of 65 million (1 in 240,000).

meaning is somewhat banal: 'Choose this recipe!' Our brains are very ready to notice connections and to ask what they mean. When those connections are personal and carry emotional content, it's hard to dismiss them as simple manifestations of blind chance, even if that is what they are. This compulsion to make a good story out of things that have simply happened to happen is the narrative fallacy, and it arises from a familiar source: the **randomness–meaning** duality.

The Texas sharpshooter, the story goes, fired six bullets from his Colt revolver into a barn wall. He then reached for a pot of paint and painted a target, centred where his bullets went. Chance events appear different when we look back at them. If we define our targets with hindsight, it's no surprise that we tend to hit them. We are all susceptible to after-the-fact analysis, looking at outcomes and mistaking coincidental alignments for meaningful patterns. If you're serious about analysing and understanding chance, you must paint the target before the shots are fired. We recognize another of our dualites: **foresight–hindsight**. We note coincidences only in retrospect, once the thing has happened. I didn't set out on that drive to the Lake District wondering about the chances of hearing a saxophone solo on the car stereo at the same time as a freight truck with the soloist's name painted on it drove past. But that's what happened.[63] Coincidences fool us because they work backwards: we try to work out the chances only after the observation has been made. It's an extreme example of survivorship bias.[64] We pay attention to the weird events that were impossible not to notice and ignore entirely the billions of potential coincidences that didn't actually happen.

Synchronicity

The urge to find explanations for coincidences can be very strong. The idea of synchronicity, developed by the analytical psychologist Carl Jung, arises from this impulse. Jung defines this concept as 'temporally coincident occurrences of acausal events'. The word acausal (without cause) is the crux here. Writing at a time when the physical theories of relativity and quantum mechanics were overturning the way in which we understood so many things about the universe,

---

[63] 'Malherbe', if you're curious. The wonderful Didier Malherbe playing with the extraordinary seventies band, Gong.

[64] Survivorship bias is, for example, when we analyse the path to success of those who have succeeded, without consideration of the many who may have made similar choices and yet failed.

Jung sought an explanation for coincidence in something detached from cause and effect:

> Therefore it cannot be a question of cause and effect, but of a falling together in time, a kind of simultaneity. Because of this quality of simultaneity, I have picked on the term 'synchronicity' to designate a hypothetical factor equal in rank to causality as a principle of explanation.

Arthur Koestler's 1972 book *The Roots of Coincidence* makes use of Jung's ideas of synchronicity, rejecting the idea of meaningless coincidence. Koestler builds a case for explaining paranormal phenomena, extrasensory perception, telepathy, and the like. The book has not stood up to critical analysis: psychologist David Marks calls this 'Koestler's Fallacy', the idea that chance cannot be responsible for uncanny alignments of random events. Perhaps something that contributes to Koestler's Fallacy is our failure to recognize all the many moments, the opportunities for coincidence, where something notable could have happened, but nothing did. Only the remarkable is remarked upon.

We know from personal experience how tempting the narrative fallacy is. We look for meaningful explanations for these improbable events: perhaps explanations that involve an external, unseen agency beyond the natural world, and beyond the blind operation of chance. But Jung's idea of synchronicity, for all its seductiveness, has not been supported by evidence, and is now widely discredited. If we say that it is impossible to imagine that this thing or that thing has happened by chance, then we must count that as a failure of imagination, not of rationality.

Life regularly throws up coincidences that a novelist would reject as unacceptable to readers. A novel must be driven by inevitability, by character, situation, and historical context, and unfold in a plausible way. Throw in a coincidence and it feels like a cheap trick. This need for a coherent narrative carries over to our own lives. When in real life we come across a coincidence that baffles our powers of explanation, it feels like a bad plot device. We are tempted to invent a hidden backstory to account for it. If we can't find an explanation that boils down to the reasonable operation of probability, over multiple opportunities, then we'll be tempted to align ourselves with Jung and Koestler, and seek meaning in a fictitious principle such as synchronicity. So, in the interests of rationality, let's see what factors might be involved in the making of a coincidence. The bones of the situation are these: given a probability for a coincidental event and a number of opportunities for that event to occur, how often should we expect such an event to happen?

What makes a coincidence?

Professor David Spiegelhalter of the University of Cambridge maintains a database of coincidences reported by visitors to the *Understanding Uncertainty* website.[65] One of the self-reported stories runs as follows:

> My father passed away 35 years ago, was buried in the local cemetery and the grave number turned out to be 9134—the same as the last four digits of our telephone number.

This is a useful example to analyse. It's simple, there are only a couple of moving parts, and it is precise—so we can easily get a handle on it.

Let's look first at the arithmetic. Breaking it down to essentials, what has happened is that one four-digit number has matched the last four digits of another number. What are the chances of that? Well, if we had specified in advance what types of number (grave number, telephone number) we would consider, and if we can treat those numbers as randomly distributed over the full range of possibilities, then it's straightforward. There are 10,000 possible last-four-digit combinations for the telephone, and so there is a 1 in 10,000 chance of a match by any given other four-digit number, such as the grave number.

First observation: A 1 in 10,000 event is a remarkable occurrence *if it happens to you*. That it should happen to someone, somewhere, sometime, is entirely unremarkable. Around half a million deaths occur in the United Kingdom every year. Around a quarter of them will result in burials, and not all (but most) of the deceased will have telephone numbers. Let's estimate that the number of opportunities for this precise kind of coincidence to happen is around 120,000 in the United Kingdom each year. We can then expect that, on average, around 12 grave–telephone number matches will happen every year in the United Kingdom.

Second observation: The story as reported matches the grave number to 'our telephone number', presumably the family's land line. Is it possible that the reporter might have been equally impressed had the grave number matched some other prominent number associated with their father? His *business* telephone number? His passport number? The *first* four digits of the telephone number? His year of birth?

This story (which, when we come to analyse it, is fairly uninteresting) is characteristic of many coincidence stories. From the individual perspective, it was

---

[65] https://understandinguncertainty.org/coincidences.

sufficiently surprising for the reporter to remember it 35 years later: from a collective perspective, the numbers show that it was entirely to be expected in a large enough population. It is a conjunction of facts noted in retrospect, a conjunction that happened to catch the reporter's attention. Perhaps there were a thousand pieces of information associated with his father's death that might have led to a noticeable coincidence, but which were near misses and went unnoticed.

Typical coincidences

Professor Spiegelhalter's website for collecting coincidences invites respondents to tag their stories by choosing one or more circumstances from the following list:

- Meeting someone you know in an unlikely place
- Finding a link with someone you meet
- Repetition of very similar events
- Simultaneous occurrence of events
- An object reappearing
- Finding a link with an object
- Parallel stories with multiple matches
- A matching number
- A matching name
- A matching date
- Matching music.

Most of these are difficult to analyse in anything like a rigorous way. The grave–telephone number story is an example of a matching number, and the numerical aspect lends itself to a mathematical approach, but even so, that's not the whole story. And when it comes to coincidences of the finding-a-link-with-an-object kind, it's hard to know where to start. To estimate the probability of, or count the opportunities for, such coincidences seems nigh impossible. Still, we can set out a general approach.

How to explain away a coincidence

If you're setting out to explore how outrageous a coincidence is, the general pattern runs like this:

- **What defines your coincidence?**
  Consider how broad or narrow your definition of a coincidental event should be, and how many events could possibly match that definition.
  (In our example, this is the matching of four digits of two numbers—not just grave numbers— that are notable around the time of someone's death—perhaps 10 possibilities?)

- **How wide is your circle of contacts?**
  Consider how many people the coincidence could happen to, that would mean you would hear of it.
  (In our example, the reporter is describing a personal experience, not a reported one; so only one.)

- **How frequent are the opportunities?**
  Consider how many times an opportunity is presented for the coincidence to occur.
  (In the United Kingdom there are around 500,000 deaths annually.[66])

- **What are the chances each time?**
  Consider the probability of the event happening to one person, once.
  (In our example, as established, 1 in 10,000.)

- **Multiply these together**
  This will give a very approximate estimate of how frequent or rare that event would be.
  (Result: a four-digit number match-up associated with a death would happen on average around 500 times annually in the United Kingdom.)

Let's go through each of these steps.

What defines your coincidence?

Opportunities for surprise are everywhere. Telephone numbers, car registration plates, faces in the crowd, names in the newspaper, music played on the radio, Twitter tweets and Facebook posts. And any conjunction of these with one another, or with memory, may trigger a moment of astonishment. The human brain is very good at spotting connections. A million non-coincidences will go unnoticed. But when a connection is made, light bulbs flash and bells

---

[66] Not only burials: we're no longer limiting our thinking to grave numbers.

ring in the brain. Coincidences arise from hindsight, and hindsight has very powerful filters.

Counting up the match-ups that could possibly trigger your mental coincidence detectors is impossible in any precise sense. But, when you notice a startling coincidence, be aware of exactly how many elements of your situation could have come into alignment and struck you just as forcefully and how many opportunities for coincidence there might be.

How wide is your circle?

How many people are you connected with? What is your exposure to tales of unlikely occurrences? Would I have been as astonished had the airplane-seat incident happened to my wife or one of my sons travelling with me? What if I had heard the interaction happening to another pair of travellers across the aisle from me? If we admit that coincidences happening to other people are as unusual as those we ourselves experience, suddenly the number of opportunities is widened hugely.

Coincidences are shared: we tell our friends and family and they tell their friends and family. Six degrees of separation, we are told, is what characterizes the network that links all of us together. Steven Strogatz and Duncan Watts in a 1998 paper, 'Collective dynamics of "small-world" networks', analysed networks that have a high degree of clustering (friends of my friends are likely to be my friends, too), but also have a small proportion of far-reaching connections (some of the people I know are in very different social circles). Labelling these networks as 'small-world' reflects the common reaction on discovering a shared acquaintance from a chance meeting, and it is a characteristic of these networks that they can result in surprisingly short chains of connection between people who you might have thought were far apart.

Robin Dunbar, an anthropologist, has determined that the largest number of people with whom we can have stable social relationships is around 150. Making some rather broad assumptions about network sizes and overlaps, the number of friends of friends (two degrees of separation) that someone has is generally around the order of 1,000 to 10,000. So when you hear the story of a coincidence that happened to a friend of a close contact, adjust your likelihood calibration by at least a factor of a thousand, and possibly more.

When it comes to social media contacts, the network is even tighter. Facebook estimates that there are on average just over three and a half degrees of separation between the over 2 billion users of that network, and that a user with

100 friends would have access to over 25,000 friends of friends.[67] LinkedIn reckons that trustworthy business contacts can be formed using chains of up to three links in length (the two people in the middle trust each other, and each trusts one of the people at the end of the chain), and that means that there are more than 50,000 trusted connections for the average user.

Any striking coincidence within your broader network has a good chance of coming to your attention. If that happens through public media, it means that everyone in the world is potentially on your radar. The most unexpected events are the ones considered most newsworthy. So it is hardly surprising that unusual occurrences seem more common than pure chance would suggest. Seven-billion-to-one odds turn into near certainties if the world's media can filter the events and direct only the most astonishing to your attention.

The population of the world is around 7.5 billion, the population of the United Kingdom is around 65 million, and your friend-of-friends network probably has around 5,000 people.

If something happened to one in a million people, how many does that mean in the world, in the United Kingdom and in your circle? What about one in a thousand?

	World	UK	Your circle
Population	7.5 billion	65 million	5,000
1 in a million	7,500	65	0
1 in a thousand	7.5 million	65,000	5

What does one in a million look like? Here are some things that apply to roughly one in a million people.

- Rare diseases are defined in the European Union as those affecting fewer than 1 in 2,000 people. (But when the prevalence of a disease is very low, the accuracy of measurement is also very poor.)
  - Creutzfeldt-Jakob disease[68] is estimated to have a prevalence of a little less than two in a million.
  - Ewing sarcoma, a cancer of bone and soft tissue, has a prevalence of about one in a million.

---

[67] If you're a member of a social network, on average your friends will have more friends than you do. This is called the friendship paradox and is brilliantly explained by Steven Strogatz in this article: https://opinionator.blogs.nytimes.com/2012/09/17/friends-you-can-count-on/?_r=0.

[68] In the 1990s in the United Kingdom, the so-called 'mad cow disease' (bovine spongiform encephalopathy) affected the brains of cattle and triggered a health scare. The fear was that the disease would cross between species and show up as Creutzfeldt-Jakob disease in humans.

- Around 5,000 people have climbed Mount Everest. That's around one in 1.5 million of all the people in the world.
- Slightly more than one in a million of the world's people lives in the small island nation of Tuvalu (population 11,100).
- The proportion of the world's population that lives in the Falkland Islands (population 3,200) is a little under one in two million.

What does one in a thousand look like? Here are some things that apply to roughly one in a thousand people.

- In mid-2018, the United Kingdom Office for National Statistics reported that some 32 million people were in employment. Among these there were around 32,000 in each of these kinds of job:
  o Furniture makers and other craft woodworkers
  o Launderers, dry cleaners, and pressers
  o Human resources administrative occupations
  o Chemical scientists
  o Floorers and wall tilers.

So if you were a human resources person, a cabinet maker, a launderer, or a chemist (as opposed to a pharmacist[69]), you would be 1 in 1,000.

- Around 65,000 people participate in the sport of archery in the United Kingdom. That's roughly 1 in 1,000 of the whole population of the United Kingdom.
- About 1 in 1,000 of the world's population lives in Germany.
- About 1 in 1,000 of the world's population lives in Thailand.

How frequent are the opportunities?

There are around 2½ billion seconds in a full human lifetime—plenty of opportunities for coincidences to happen. If we say that a 'lifetime' is 80 years and a 'working lifetime' is 50 years of 250 days of 8 hours, we can calculate how many chances there are for coincidences to occur. For example, if there was a lottery with a one-in-a-million chance and you could play it every minute of every working day, then on average you would win about six times in a working lifetime.

---

[69] There were 66,000 pharmacists, around 1 in 480 of all employed people. But if you take as your base the number of all people in the United Kingdom (including children, the elderly, and the unemployed), then pharmacists are indeed about 1 in 1,000.

How many events on average if the chance is...?	How many in a lifetime?		How many in a working lifetime?	
	Opportunity every day	Opportunity every second	Opportunity every day	Opportunity every second
1 in a million	0	2,500	0	360
1 in a thousand	29	2,500,000	13	360,000

How many events on average if the chance is...?	How many in a year?		How many in a day?	
	Opportunity every day	Opportunity every second	Opportunity every day	Opportunity every second
1 in a million	0	31	0	0
1 in a thousand	0	31,000	0	86

This is no more than a multiplication table converting the proportions of one in a million and one in a thousand to different stretches of time, expressed as days and seconds. It does show, though, that unless opportunities happen very frequently, on the scale of seconds, a one-in-a-million chance is vanishingly small from the viewpoint of a single human's experience. On the other hand, one-in-a-thousand chances can be pretty ordinary.

What sort of events happen with a frequency (to an order of magnitude) of one in a million times?

- Plane crashes: 2018 was a bad year for commercial aviation. The proportion of flights crashing with fatalities was around 1 in 3 million. The five-year average for 2014–2018 was around 1 in 4 million. Crashes without fatalities average a little under 4 per million flights.
- Winning single-number bets in roulette four times in a row is about a 1 in 2 million chance.
- In the United Kingdom, every day you run a risk of around one in a million of dying from non-natural causes.

Here are some things that happen with approximately 1 in a 1,000 probability.

- Your chance of winning in a 'numbers game' lottery based on matching three-digit numbers is exactly 1 in 1,000 (if the game is fair).
- Sharing an exact date of birth with someone born within 18 months of you is about 1 in 1,100.

- Winning single-number bets in roulette twice in a row is about a 1 in 1,400 chance.
- Being dealt a full house in five-card poker happens about 1 in 700 times.

What are the chances each time?

What are the chances of matching numbers?

On the face of it, working out the chance that two numbers will match should be relatively straightforward. A four-digit number has 10,000 possibilities; so the chance of your bank card's PIN matching that of your friend is 1 in 10,000. But not all memorable numbers are chosen evenly from all possibilities. Many identifying numbers used in computer systems have some structure, and may incorporate error-correcting elements, all of which reduce the variation and increase the chance of a match. Although you might think that a 16-digit credit card number would have 10^{16} possible values, it does not. The first four to six digits are the issuer identification number and the last is a check digit mathematically calculated from the first 15. This makes the number of possibilities considerably fewer.

What are the chances of matching dates?

The 'birthday paradox' is well known, but worth repeating. How many people must be gathered together for it to be more likely than not that two of them share a birthday? The answer is surprisingly small: 23. The reason for this lies once again in big numbers.

If 23 people are present, there are $\frac{22 \times 22}{2} = 253$ pairs of guests, and each pair has a chance of hitting our match-up criterion. The two people involved will no doubt be mildly astonished. Each will think: the chance of 1 in 22 people sharing *my* birthday is rather unlikely—and they will be correct to think that. But detach the coincidence from any specific person, and the chances of any match-up occurring becomes much less surprising.

The birthday paradox encapsulates a surprising conclusion in an elegant and understandable way. And it's not just restricted to a gathering of a couple of dozen people. Think of a small family. Perhaps, out of four grandparents, there are two who are still alive. Then there are Mum and Dad and two children. Let's extend the family with two sets of aunts and uncles, each of those couples with two children, cousins to our core family. That makes 16 individuals with, among them, 16 birthdays, 5 wedding anniversaries, and 2 anniversaries of death of the missing grandparents. In all, that makes 23 notable dates, and that matches the birthday paradox: it's even money that two of those notable dates will align, just in that small family.

What are the chances of matching names?

Two young people fall in love and agree to marry. The day comes when their respective parents meet for the first time. The groom's parents are called George and Helen Willis, and they are introduced to the bride-to-be's parents, called…George and Helen Willis. What are the chances of that?[70]

Three things need to match up for this coincidence to happen: the surnames of the two families, the fathers' first names, and the mothers' first names. 'Willis' comes around 200 in the list of the most common surnames in England and Wales: about 1 in 2,000 people have that name. George was the twenty-fifth most popular boy's name in England and Wales in the 1950s: around 1 in 110 boys. Helen was less common, just 1 in 340 girls in that decade was named Helen. So we might say that the chances of one person called Willis meeting a potential partner with the same name is 1 in 2,000, the chances of encountering a George (of any surname) of the appropriate age is 1 in 110, and for a Helen it is 1 in 340. Multiply them together to get a 1 in 75 million chance. That's less likely than winning the National Lottery.

But hold on. Can this be right? We are assuming that just because the proportion of Willises is 1 in 2,000, the probability of a random encounter between Willises is 1 in 2,000. The population of England and Wales is not like a well-shuffled deck of cards. There are regions where certain surnames are more likely to be found. Here is the distribution of Willises in the United Kingdom:[71]

So a Willis from England is much more likely to be found near Kingston Upon Hull (the East coast hotspot visible on the map) or Oxford (the one west of London) than in other locations. And this must mean that one Willis is more likely to meet another than crude national statistics would suggest. Maybe we're out by a factor of four, and the chance of a Willis finding another Willis is more like 1 in 500.

Now think about the first names. Are the chances of Willis as a surname and George as a first name really independent? Issues of class or ethnicity could

---

[70] The story is real, the names are not. The made-up names I've used here have been chosen to be about as common as the ones in the true story.

[71] Thanks to http://named.publicprofiler.org/. 'Willis' is the name I have substituted for the real one in this story. The actual name involved was even more patchy in its distribution.

mean that a Willis would have a first name picked from a narrower set of names than the broader population. Perhaps Georges have a similarly clumpy geographic distribution that overlaps with the Willis distribution? What are the chances of a man being named George, given that he has the surname Willis? For similar reasons (geography, class, ethnicity), perhaps Georges hook up with Helens more than would be suggested by simple proportions based on national statistics.

Published statistics are not designed to answer these questions, so evidence is hard to find. We'll make some heroic assumptions. What if the chance of a male Willis of the right age being a George is twice as likely as raw statistics suggest? 1 in 55 rather than 1 in 110. What if the chance of a George and a Helen finding each other is also twice as likely? 1 in 170. Then the combined chance of these name matches becomes more like 1 in 5 million.

What's the appropriate measure of exposure here? The number of marriages? Around 250,000 happen in England and Wales each year, which means there is a 1 in 20 chance that such a match-up of families would happen in any year. Our reasoning here has been very approximate, and it shows how complex it can be to reason constructively about probabilities in the real world, but perhaps it also shows that what seems like an outrageous coincidence is really not all that improbable.

Multiply these together

When you bring together a broad definition of what coincidences might look like, a wide circle of friends of friends from whom reports might reach you, and frequent opportunities for them to happen, you can end up with a very big number of opportunities for coincidence to come to your attention. Even combining that with a very small chance of occurrence can easily leave you with an answer that is reassuringly within the bounds of plausibility. Of course, it's a lot of effort to go to, and you're not going to reach for your calculator when you meet Alan from primary school in a bar in Buenos Aires and he asks 'what are the chances?', but in the back of your mind you might have the answer: 'it's not impossible'.

The inevitability of miracles

The Law of Truly Large Numbers: With a large enough sample, any outrageous thing is likely to happen.

Diaconis and Mosteller (1989)

Cambridge maths professor John Littlewood, in playful mode, defined a 'miracle' as an event that had a one-in-a-million chance of happening. He then reckoned that one second was enough to notice such an event and that, assuming you were sufficiently alert for eight hours each day, it would take on average around 35 days (amounting to 1,008,000 alert seconds) for a miraculous event to happen to each of us. So, each of us should experience more than 10 miracles a year.

Coincidences impress themselves on our minds very forcefully. They are amplified by the way our brains crave explanation. Many people, and not just cranks, believe that there is more to coincidence than pure chance. We have found good scientific and natural explanations for many phenomena that in previous ages would have been attributed to supernatural agencies, but when it comes to genuinely random occurrences, particularly where there is a personal connection, an astonishing coincidence can feel like something that violates our understanding of how things work, something that demands explanation. We instinctively reject the idea that the seemingly impossible is in fact quite probable. There is then a risk that we revert to the irrational and the magical, to the detriment of clear thinking. For some things really do happen without logical reason, without synchronicity or other hidden mechanisms, but just randomly.

Littlewood's playful suggestion that we would each experience a miraculous event about once every 35 days is equivalent to saying that there is a 2.86 per cent chance that any given day will contain a miracle. So what is the chance that an entire year would pass without a single miraculous day? About 0.0025 per cent, around 1 in 39,000. A year without coincidences? That really would be remarkable.

Notable chances: Littlewood's miracles

Littlewood suggests the probability of a miracle happening on a given day is:
Probability: 0.0286
Percentage: 2.86%
Rounded-off proportion: 1 in 35

Approximate fair betting odds: 34 to 1

Representative example:
A lifetime: 80 years (the rows) of 365 days each (the columns) = 29,200 days
Million-to-one miracles in this example: 861 = 2.95% (expected average: 834.3 = 2.86%)

And in *this* example, there are no years without miracles. The year with the fewest has five of them.

Comments on: Signs and portents

The story of Mark Twain recognizing that his death was aligned with successive appearances of Halley's comet is one worth telling, not least for the charmingly self-deprecating way in which he wrote about it. But is it really that extraordinary?

If we start at the beginning, there is no hint of significance about the alignment of the date of his *birth* with the comet's appearance. Plenty of babies were born around that time, and indeed, his date of birth was two weeks later than the perihelion; so the alignment was not even as precise as implied.

Then, 75½ years is about as long as a man of his time could expect to live. Writing up the tale of his long life, having lived 74 years, he could surely not have expected to be around for very much longer. Perhaps he knew his body was failing, and no doubt he had carried this idea, that his life was paired with the comet's cycle, with him all his life.

We know Mark Twain mostly, these days, as a writer. But he was also a performer, delivering hundreds of readings and lectures. And any stand-up performer, as perhaps any writer, knows the importance of timing, and how to deliver a punchline to a gag that had been set up earlier. I can easily imagine that, having held out long enough to make his rendezvous with his cometary twin, the old showman knew exactly when to leave the stage.

The third duality: Foresight–Hindsight

If two roads diverge in a wood, only one can be taken, and that can make all the difference. Multiply that divergence many times over, perhaps with some choices made at random, and look where we are, here and now, at this singular place and time. Who could have predicted it? Looking back, though, we can easily track the route we took, and explain our choices. Those choices, those branchings, have made all the difference.

So this is the third duality that makes it hard to think about uncertainty. Even if we know all the possibilities and how to compute their probabilities, the world changes when the dice settles, and comes up five, and the game is won. The probability function collapses, what was potential becomes actual, and the consequences follow. When the battle has been won, it's easy to claim that victory was inevitable. Hindsight has all the explanations.

Hindsight is also selective. The paths not taken get barely a second glance. The successful entrepreneur stands on stage and talks of hard work and late nights, and exhorts the audience to keep going, to push through the hard times, as he did, because *that* is the route to success. The nine out of ten who worked just as hard, just as late into the night, but nonetheless failed to achieve their goals have no audience for their stories.

It's said that we fail to learn from history. That may be right, but perhaps the reverse is also true. A terrorist attempts to set off a shoe bomb, and so we start having to remove our shoes in airport security. We say we won't be fooled again. But the next attack, using a different weapon, takes us by surprise. Who could have foreseen it? What were the chances? Looking back, it's easy to pick out a plausible narrative. This happened, and then that happened. One thing caused another. We should have predicted it. We should have known.

Psychologist Norbert Kerr wrote about the dangers of using hindsight in research, coining the term HARKing: Hypothesizing After the Results are Known. The overambitious researcher dredges through their data to find trends they never anticipated when starting the project. Maybe those findings are genuine, or maybe they are simply the result of chance, but by attacking the problem with hindsight, the research is compromised. The researcher has painted the target after the shots have been fired.

Hindsight highlights what did happen at the expense of what might have happened and creates a sometimes spurious logic to explain how we got to where we are. We should always be careful of assuming that the same logic is a sound guide to the future.

CHAPTER 6

What Luck!

Luck is statistics taken personally.

Penn Jillette

How long can your luck last?

'Normally,' says the mystery man at the bar, 'the joker in the pack is a lucky card. Not in this game. Here's a standard pack of 52 cards. I'll shuffle a joker into it. Here, cut the deck a few times so you're sure I haven't stacked it in my favour.'

He hands me the deck. I lift off about a third of the cards, replacing them under the others.

'Now,' he says, 'you deal one card at a time, and count them off as you go. If you can deal half the pack before that unlucky joker shows his face, you win. Ten dollar stakes.'

'Half the pack? Do we take that as 26 cards or 27?'

'I'll be generous. If you survive the turn of 26 cards, you win. If the joker shows before then, you're "dead", and I win. Okay?'

You agree. The odds seem to be marginally in your favour; so you start dealing cards one at a time. Six cards in, no joker yet, and the stranger's phone rings. You pause while he takes the call.

'I am so sorry, I really need to go right away. Another time?' and he moves to gather up the cards.

'Hold on. Don't you owe me something?'

'Do I?'

'Yes you do. I've survived six card flips now, and so the chance of me winning must be a bit better than 50/50. It wouldn't be fair for you to abandon the game now.'

What are the Chances of That? How to think about uncertainty. Andrew C. A. Elliott,
Oxford University Press. © Andrew C. A. Elliott 2021. DOI: 10.1093/oso/9780198869023.003.0006

'Hmm, you have a point. Ok, how much compensation do you think you deserve?'

Do you have an answer? Discussion at the end of the chapter.

O Lucky Man!

Frane Selak, a Croatian music teacher, has been called the world's luckiest person. The story of his luck includes seven (lucky seven!) occasions where he came very close to death. The incidents involved a train, a plane, a truck, buses on two separate occasions, and two cases where he himself was driving and his car caught fire. These were not mere near misses: in several of these accidents other people were reportedly killed. Then in 2003, at the age of 73, Selak won the lottery, scooping a prize of €800,000 (about a million dollars).

Perhaps. Some doubt has been expressed about these claims: the train crash and the plane crash may not have happened quite as recounted. One of the bus incidents appears to have been relatively minor. But let's not quibble. If not Frane Selak, there is undoubtedly somewhere a person whom we would call unusually lucky, who has repeatedly survived near-fatal accidents, and who has also had a slice of good fortune along the way. And that all makes for a good 'news' story. In all the lives of 7.5 billion people alive on Earth, there will be some for whom the dice have rolled the right way, against the odds, far more often than not. In hindsight, and possibly with a little light editing of the story, we can tell the tale of the world's luckiest person. But what is luck?

The wheel of fortune

O Fortune / like the moon /
you are changeable
ever waxing / ever waning /
hateful life
first oppresses / and then soothes /
playing mind games
poverty / and power /
it melts them like ice.

O Fortuna / velut luna /
statu variabilis
semper crescis / aut decrescis /
vita detestabilis
nunc obdurat / et tunc curat /
ludo mentis aciem
egestatem / potestatem /
dissolvit ut glaciem

From Carl Orff, *Carmina Burana*

When as a child you lost too often at *Snakes and Ladders*, you may have sulked, and charged your opponent with being 'too lucky'. We know this to be a childish attitude, that a run of wins is nothing more than a random outcome. But when things repeatedly go against you, it feels unfair. When chances appear to fall consistently in the same way, for good or bad, we attribute it to luck. It seems as if luck is inclined to stick to a person, or to a place, or to a garment, or to the lucky charm in your pocket. While coincidence jumps out of the bushes and startles you, luck is a creature that forms an attachment and follows you home.

We expect and accept that chance will interfere with our plans, and may push them off course. But if we lose on the swings, we hope that the roundabouts will compensate. If the run of luck goes against us too many times, it feels unfair. We expect natural justice to establish cause-and-effect links between actions and outcomes. Hard work and virtue should be rewarded while idleness and transgression deserve punishment. That's fair. And if the wheel of fate turns against us temporarily, it seems only fair to expect it to keep turning and in time restore the balance. When things are going too smoothly, we get nervous that the run of good luck will soon end. But chance is not a wheel. There is no tendency to turn things around. The most we can expect is that the past will not affect the future. Extended runs of good and bad fortune can and will happen, but not necessarily in alternation.

When chance breaks the chain of cause and consequence, we put it down to luck. If a person dies at too young an age, a random victim of disease or accident, we try to make sense of it. We invoke superstitions, or fill our minds with regrets about inaction or ill-judged actions. When a lottery win confers unearned riches, the winner thanks her lucky stars. She knew those numbers— her wedding date and the registration number of her first car—would come good sooner or later.

Luck becomes a thing in itself, and superstition is how we interact with it. Superstition, though akin to religious belief, seems more primitive, more deepseated. So we blow on the dice before rolling them, invoking some form of raw magic. We glance at the heavens before taking that crucial putt on the golf green. We avoid walking under ladders or having 13 people seated at the table. We invoke the assistance of Ganesha or St Christopher. We cross our fingers and knock on wood.

If it's so easy to attach the label 'lucky' to someone for whom the chances always seem to fall favourably, then it's not too hard to entertain the idea of predestination, that someone's path through life is laid out in advance. The

Fates of the ancient Greeks were incarnations of destiny, and were three in number: Clotho, spinning the thread of life, Lachesis, measuring out the length of that thread, and Atropos, cutting off that thread to the measured length. Collectively, the Greeks called them the *Moirai*, the Apportioners, delivering to each person their allotted portion. That's your lot.

Belief in fate is one way to reconcile ourselves to the unfairness of life. Blind chance is hard to accept, and explanations based on notions of punishment and reward can leave us filled with guilt when bad things happen. But the idea of fate, that in some unseen realm the course of our lives has already been laid out, provides a kind of explanation that absolves us of guilt and responsibility. The bullet had his name on it. There was nothing that could be done. What bad luck.

Fortune-telling

It's only a short step from believing that there is a hidden plan for our lives, to wanting to get a peek at the map. Where can we find the clues to reveal what would otherwise remain hidden? Is the length of your life written in the creases of your palm, or the bumps of your skull? Or should we instead look at the tea leaves? Oddly enough, one of the ways that people have sought to glimpse this supposed occult information is through techniques that rely on randomness. Surely, the reasoning goes, the answers to such mysteries can only lie in the things for which we have no other explanation, things that would otherwise have no reason to be as they are. Where else could those secrets be hidden but in chance occurrences?

In ancient Rome an augur, a specialist priest, might be consulted before important decisions were made. He would watch the seemingly random movement of birds against a designated sector of the sky, using established rules to interpret their flights. Would the proposed venture meet with heavenly approval or punishment? Such a bird-watching priest was an *auspex*, literally 'an observer of birds'. A good reading would signify that the occasion was 'auspicious'. On other occasions a *haruspex* would be called for, one who specialized in examining and interpreting the internal organs of sacrificed animals spilled out chaotically on the temple floor.

Over the centuries, countless ways have been devised for attempting to use randomness to gain insight into those sought-after plans that are shielded from our sight. Here are just a few:

- **Bibliomancy—books**
  Place a book with its spine on the table and allow it to fall open. Pick a passage with closed eyes. Interpret the passage. Religious scriptures are often used, as sometimes are books of poetry, which offer great scope for interpretation.

- **Cartomancy—cards**
  From a shuffled deck, cards are laid out in a formulaic way. Each card has its significance, influenced by position in the layout and its relation to other cards. Often the Tarot deck is used, which contains 78 cards, including four suits of 14 cards each, plus 22 Major Arcana cards with names like The Magician, The Wheel of Fortune, The Empress, and The Hanged Man. Plenty of scope for creative interpretation there!

- **Cleromancy—random mechanisms in general**
  This covers many ways of fortune-telling using devices specifically made for the purpose of randomization. It includes the rolling of dice, the tossing of coins, and also the 'casting of lots'.

- **Literomancy—letters, characters**
  The literomancer interprets the way in which a client has written a single word, a kind of handwriting analysis. In Chinese culture, a literomancer may interpret a single written character.

- **Lithomancy—stones, reflections from stones**
  A collection of visually distinct stones is assembled (13 is a typical number). To these stones are assigned specific associations such as the Moon, Water, the Universe. A number of stones are drawn randomly, and an interpretation of the associations is made.

- **Rhabdomancy—sticks, staffs, rods**
  Any divination technique involving the use of straight sticks; for example, interpreting the patterns made when a bundle of sticks is held upright and then released. Traditional dowsing for water is a form of rhabdomancy.

Common to all these techniques is that they provide an opportunity for randomness to show itself, which forms the basis of the fortune-telling. No doubt the real skill of the successful practitioner lies in a psychological reading of the client, but their interpretation must be consistent both with the random outcome and with their understanding of the needs of the questioner.

Consulting a fortune-teller may be an opportunity for a conversation between a client with a problem and a skilled counsellor, but the random aspect of the divination plays an important role. It creates mystery and thus establishes an air of authority in the reading. It is also a source of inspiration to the interpreter, who must be an improviser, responding in the moment to a conjunction of chance and situation. In many of these ways of telling fortunes, the random element is not amenable to easy analysis. But some are more mathematical. One such is the *I Ching*.

Cleromancy: the *I Ching*

The Philip K. Dick book *The Man in the High Castle* is an alternative-history novel, where the Axis powers, notably Germany and Japan, were victorious in the Second World War. The west coast of the United States is under the sway of the Japanese, and has in many ways been influenced by Asian culture. Several characters in the book make time regularly through the day to consult the *I Ching*, the ancient Asian oracle, to obtain guidance for action.

Published in the late ninth century BCE, the *I Ching* or *Book of Changes* is a Chinese work of divination. A randomly generated figure (a hexagram) serves as an index to the written text. Dick's characters from time to time cast hexagrams and consult the oracle, and so allow chance to have a part in directing their actions. The unexplainable, the random, becomes a crack in their everyday world through which new information can enter and guide their choices. One reading of *The Man in the High Castle* (which is signposted by the explicit incorporation of the *I Ching*) is that this is a model for history itself. The alternative world it describes has resulted from small deviations from the history of our world. Small, seemingly random, nudges have led to huge consequences.[72] This was chaos theory before chaos theory became trendy.

It came as no surprise to me to learn that Philip K. Dick had in fact used the *I Ching* as part of the process of writing his book. For even if we reject the idea that the *I Ching* is some sort of portalgranting access to occult wisdom, it is certainly true that random inputs can be useful in decision-making. Committing to a definite course is often preferable to the paralysis of indecision.

The *I Ching* is also an example of probabilities that may be unfamiliar to some readers. Casting the *I Ching* involves the random generation of hexagrams, some examples of which are shown here. Each one consists of six horizontal lines, which may be broken (*yin*) or not broken (*yang*).

---

[72] And in *The Man in the High Castle*, the characters read a book in which an alternative history to their reality is described. In that world, the Allies won the war, but that alternative alternative world is by no means the same as our world. Different choices, different outcomes.

35	36	37	38	39
䷢	䷣	䷤	䷥	䷦
'jin' Prospering	'ming yi' Darkening of the light	'jiā rén' Dwelling people	'kuí' Polarising	'jiǎn' Limping
43	44	45	46	47
䷪	䷫	䷬	䷭	䷮
'guài' Displacement	'gòu' Coupling	'cuì' Clustering	'shēng' Ascending	'kùn' Confining
51	52	53	54	55
䷲	䷳	䷴	䷵	䷶
'zhèn' Shake	'gèn' Bound	'jiàn' Infiltrating	'guī mèi' Converting the maiden	'fēng' Abounding

As each of the six lines can occur in one of two main variations, it follows that there are 64 (2⁶) possible hexagrams. To consult the oracle, start with a question in mind. Then cast a hexagram, using a random process. That is then used as an index into the *Book of Changes*, which contains texts of various kinds for each hexagram, which are interpreted in the light of your question.

The hexagram can be cast in several ways. The simplest involves the throwing of three coins to determine each line in the shape.[73] Heads are given a value of three points, tails two. The total point value from the first throw determines the bottom line of the hexagram. A total of 6 (which can only come from three tails and so has ⅛ probability) or 8 (two heads and a tail: ⅜ probability) signifies yin, 7 (two tails and a head: ⅜ probability) or 9 (three heads: ⅛ probability) signifies yang.[74] So yin and yang are equally likely, but as you will see, this is not the end of the story. A further five throws of the three coins are made to complete the six lines of the hexagram.

The lines in the hexagram are further characterized as 'moving' (also called 'old') or 'static' (also called 'young'). Point totals of 6 and 9—which are less

---

[73] We've looked at the chances for tossing three coins before, in the chapter on probability distributions (p. 70).

[74] You'll have noticed that this is hardly an efficient calculation. Easier, surely, to count the heads and base the categorization on counts of 0, 1, 2, or 3 heads? The use of values of 2 or 3 is for consistency with the older yarrow stalk casting method.

likely, having only one possible way to arise—signify old/moving lines. Young/static lines are created for point totals of 7 and 8.

Once you've looked up the hexagram in the book and made an interpretation, you then take account of the moving lines. Any moving lines in the hexagram are flipped between yin and yang to generate another hexagram, which is also used as an alternative index into the book, for a secondary reading. The moving lines themselves can be used to find yet further readings from the oracle. If we account for the old/young distinction and the second hexagram that can be generated, there are 4,096 possible combinations.[75] This gives plenty of scope for variety and creativity in interpretation.

An older way of casting the hexagram involves a ritual based on the use of yarrow stalks, dried plant stems. A set of 50 stalks is manipulated in the hands, the caster gathering and passing handfuls to and fro in a random and ritualized way. This method, too, produces a series of random numbers between 6 and 9. When the yarrow stalks are used, the probabilities are not identical to those found when using coins: the more traditional ritual favours old yang (with $^3/_{16}$ probability) and young yin (with $^7/_{16}$ probability) over old yin (with $^1/_{16}$ probability) and young yang (with $^5/_{16}$ probability).

Yin and yang still occur with equal probability, but because the probabilities of old/moving lines and young/static lines are different, the calculation of the probability of different combinations of hexagrams is complex.

As I write this, I have just used an online *I Ching* website to cast a fortune for myself. The question in my mind was about the wisdom of continuing with the writing of this book, and the hexagram I cast was number 45 (referring to the chart shown). That hexagram is called 'Clustering', and the text I am offered is this:

The Lake rises by welcoming and receiving Earth's waters:
The King approaches his temple.
It is wise to seek audience with him there.
Success follows this course.
Making an offering will seal your good fortune.
A goal will be realised now.

[75] There are four possible states for each line, and that makes for 4^6 (4,096) possible combinations. But some of those would have no moving lines, so there would be no second hexagram. Put another way, the scheme of moving lines means that every possible pair of hexagrams has some possibility of appearing (but not an equal probability), and in some of those cases the initial hexagram pairs with itself.

Of course I know that the selection of this passage is purely random, but I am not immune to the instinct to seek meaning even in what I know intellectually to be meaningless chance. Just as I would not discard, unread, the motto in a fortune cookie, I somehow feel that the words of the oracle at least deserve cursory attention, however lightly I will take them. So, irrationally, I am pleased to read 'success follows this course'. I'm not sure what offering I should be making when I seek audience with the King, but for now I'll keep on writing, and trust that my 'goal will be realised'. Here's a twist: if you're reading this having bought this book, then that fortune has come true. That's survivorship bias, right there. It's the **foresight–hindsight** duality.

The paradox of fairness

We may no longer think of divination as a mainstream practice, but we still hang onto it in watered-down ways. In October 2011 Monica Duca Widmer and Marco Romano were candidates for the National Council in Switzerland, in the canton of Ticino. They each received precisely the same number of votes: 23,979. The case went to the Swiss Federal Supreme Court, and it was decided that the Ticino Cantonal Council should allocate the seat by drawing lots. On 25 November 2011 Marco Romano was declared the winner.

Where a decision is necessary, and yet there is no rational basis for making the choice one way or another, we are content to leave the matter to chance. In formal language, this is known as sortition, and it's long been a part of the mechanics of government. In Renaissance Florence the governing council, the Signoria, was chosen by sortition every two months. The names of those eligible were put into a bag and nine names drawn out. Whether this was always truly random, we might have some doubt. When, in 1969, the United States drafted soldiers to fight in Vietnam, they chose birth dates randomly: the first wave of draftees were all born on 14 September. And in the United Kingdom we still make use of random selection as part of the process of choosing a jury.

The flip of the coin chooses who takes the first kick of the football match. Straws are drawn and one squaddie is tasked with delivering a message through a hail of enemy fire because he has the short straw. Sortition allows a choice to be made without any individual needing to take responsibility for choosing. It's a rather disturbing paradox: we seek to make a decision in a way that is seen to be absolutely fair, and we achieve that through a mechanism that is entirely, maximally, unfair. But because the process is random, it is without bias. Since no human agency lies behind the decision, there is no one to blame.

If fate means that your life is apportioned, spun out, measured and cut off in a predestined way, the similar idea of a 'lot' has fewer negative connotations.

lot (n.)

> 'object used to determine someone's share'. Also 'what falls to a person by lot'. From
>
> *hlot* (Old English). From
>
> **khlutom* (Proto-Germanic)

The Bible has Roman soldiers 'drawing lots', gambling for the tunic of Jesus of Nazareth at the foot of the cross, making a random allocation rather than sharing the garment by tearing it into pieces. It's not the only biblical reference to random lots: Jonah embarks from Jaffa on a boat to Tarshish, and once the journey is underway a storm springs up. The sailors decide that the storm is supernatural in origin and seek to know who has brought it upon them, and so Jonah is identified as the bringer of bad luck through the drawing of lots. And when, after the departure of Judas, the Apostles find themselves a person short, his replacement, Matthias, is selected by the drawing of lots.

The word lot encapsulates the paradox of fairness. Taking one meaning, it represents a fair share. Taking another, a random outcome. Perhaps the common thread is fairness arising from a lack of bias. It's a word that has worked its way deep into our language. It's an item at auction, it's a parcel of land, it's an imprecise, but large number of things. The word is hiding in 'allot' and 'allotment'. And of course in 'lottery', a scheme where the outcome is determined by lots.

Making your own luck

Were it not for his lottery win, Frane Selak (the world's luckiest person) might with some justification call himself the world's *unluckiest* person. He reports that his friends were unwilling to travel with him, fearing that he was jinxed. It's a fair way to look at things and suggests that perhaps attitude plays a role in how we define luck. After all, luck, whether good or bad, is a matter of how we interpret the chance occurrences of our lives.

Psychology professor Richard Wiseman has made a study of luck, specifically of people who describe themselves as lucky or unlucky. He writes: 'Although lucky and unlucky people have almost no insight into the real causes of their good and bad luck, their thoughts and behavior are responsible for much of their fortune.' He identified four basic principles at work:

- Lucky people are skilled at creating and noticing chance opportunities, whereas unlucky people tend to miss seeing them, perhaps because (as Wiseman's testing showed) they are more anxious. He found also that lucky people would often deliberately introduce variety and change into their lives, opening themselves to opportunities. The unlucky people were more likely to stick to routine, reducing the chance of fortuitous encounters.

- Lucky people make lucky decisions by listening to their intuition. Intuition can represent a distillation of years of experience, and lucky people are more likely to trust this hidden knowledge. It's not to say you should always trust your gut, but lucky people use their intuition as a prompt to examine why they have a good or bad feeling about a situation. Wiseman found that they also actively boost their intuitive abilities by keeping a clear head through activities like meditation.

- Lucky people create self-fulfilling prophecies via positive expectations. For example, when times get tough, people who think of themselves as lucky will persist in the face of obstacles, whereas unlucky people will more readily give up.

- Lucky people adopt a resilient attitude that transforms bad luck into good. Like Frane Selak who had the misfortune to find himself in life-or-death situations seven times, they focus on the positive: he survived. When misfortune strikes, lucky people say 'it could have been worse'.

Wiseman concluded that luck and chance are very different things. What we call luck is largely the consequence of our own thinking and behaviour as we respond to the chance events that surround us. We think of luck as persistent, and that may be because it's a reflection of our attitudes to the snakes and ladders we encounter, and because these attitudes form early and seldom change through life.

The idea of luck is not that helpful if we're searching for the meaning behind chance, but it remains very deeply embedded in our cultures and our language, and so also in our habits of thought. And it persists as an idea most strongly, perhaps, in the world of gambling, which is where we are headed next.

Answer to:
How long can your luck last?

The rules of the game are that you need to survive 26 card flips without the joker showing up. At the outset the joker could be in any position in the deck: positions 1 to 53. The process of dealing one card at a time doesn't change the position of the joker: if it starts off in one of the positions 1 to 26 you will lose, but if it is in a position between 27 and 53, you'll win. So the chance of winning at the outset is 27/53, or a little more than 50 per cent (50.94 per cent).

When the game is abandoned, has the situation changed at all? Yes. Since the joker was not in any of the first six cards, it must be in one of the positions remaining, 7 to 53, with equal probability for each of those 47 positions. It's still the case that you would have won if the joker was in positions 27 to 53, and that still represents 27 chances, but there are only 20 chances left to lose. So the chance of winning after six safe cards have been dealt is 27/47, or 57.45 per cent.

Your percentage chance of winning has increased, but what would be a fair settlement for abandoning the game? What is your position worth? Twenty-seven winning chances offset by 20 losing chances means a net advantage of 7 out of 47 chances, or 14.89 per cent. Since the bet was for $10, then $1.49 would seem to be the fair settlement.

But hold on. Even before you started playing, you had a small advantage: 1 chance in 53. What was that worth? That comes to 1.89 per cent or, for a bet of $10, $0.19. So perhaps, the stranger might argue, the abandoning of the game should only cost him $1.30?

CHAPTER 7

Taking a Gamble

The greatest advantage in gambling lies in not playing at all.

Gerolamo Cardano, *Liber de ludo aleae*

How long have you got?

Not again, you think, as the well-dressed stranger approaches you at the bar.

'You're a sporting person,' he says. 'I've got a small wager that might amuse you. I promise you won't lose more than a handful of pennies, but it might amuse you.'

Well, you have some time to kill, so you agree.

'Okay, how many penny coins do you have in your wallet?'

'Five.'

'That'll do. We'll take it in turns to roll three dice, and if a total of 10 comes up, you win. I give you a penny which you can add to your stack. It doesn't matter who rolls the dice, and I can guarantee that they're fair. If 9 comes up, I win. You give me one from your stack of pennies. If it's something other than 9 or 10, we roll again. We play until you're out of pennies.'

You're sure there's a catch, but where? You know by now, as Galileo knew, that a total of 10 is slightly more likely to come up than 9 when 3 dice are rolled, so the odds are definitely in your favour. But why would the stranger offer to play this game if he is likely to lose?

Is there a catch? Why did the clock stop ticking? And is there a whiff of sulphur in the air?

Answer at the end of the chapter.

What are the Chances of That? How to think about uncertainty. Andrew C. A. Elliott,
Oxford University Press. © Andrew C. A. Elliott 2021. DOI: 10.1093/oso/9780198869023.003.0007

The Gambler

The Chaco Canyon in the state of New Mexico is the site of the most extensive ancient ruins in the United States. Local oral histories tell of a powerful figure called The Gambler. He came from the south and, displaying great skill, began to gamble with the locals for ever-increasing stakes. Eventually, they gambled themselves away and ended up enslaved to him. At his bidding, they constructed the buildings that now fill the canyon. That's how the legend runs. The reality is that there is indeed extensive archaeological evidence for gambling in the Chaco Canyon. Among the artefacts found there are almost 500 that have been identified as being associated with gambling, including several different kinds of dice and playing pieces.

Greek mythology tells us how Zeus, Poseidon, and Hades gambled to decide who ruled where. They agreed to draw lots: Zeus was triumphant and claimed the Sky, Poseidon took the Sea, and Hades was left with the Underworld. If the Greek gods were gamblers, so too were the Greeks themselves. There is a reference to betting in Homer's *Iliad* where Ideomenus and Ajax make a bet for 'a tripod or a cauldron' on a chariot race that is being run as a part of the funeral games for Patroclus.

I'd wager that people have been betting against each other from the time that they had possessions that they could call their own. Items used for gambling show up in archaeological digs throughout the world. From the earliest times, people have rolled the bones, and staked their shirts, all for the thrill of flirting with chance.

The word itself?

gamble (v.)

> 'to risk something of value on a game of chance'. From
>
> ***gammlen*** (a dialectal survival of Middle English), a variant of ***gamenen*** 'to play, jest, be merry'. From
>
> > ***gamenian*** (Old English) 'to play, joke, pun'

Laws regulating gambling generally distinguish games of skill from games of chance, but the distinction is not always clear-cut. There are pure gambling games, like roulette, where the element of chance is the deciding factor that determines a win or loss. But in games such as backgammon or poker, the chance element enters in the course of play, and the player must exercise skill in responding to whatever luck arrives. So these are games that balance the elements of chance and skill.

This book is not a guide to gambling (see John Scarne for that!); nor could it ever include an exhaustive review of the many ways that people have chosen to play games, place bets, and flirt with chance. But gambling does illustrate some important aspects of the way we understand and think about chance, and that's the focus of this chapter.

It's all a lottery

A lottery is gambling at its simplest. You buy a ticket, selecting a set of numbers. All the money paid for tickets is pooled, and a proportion of the takings is returned to lucky winners if their selected numbers match the numbers that have been chosen randomly. The drawing of numbers is usually done publicly, to ensure that the lottery is fair, and seen to be fair.

Lotteries have been around for a long time. The Great Wall of China was constructed over the course of centuries, starting from the seventh century BCE. Legend has it that during the Han Dynasty in around 200 BCE the Emperor Cheung raised funds to extend and repair the Great Wall by arranging a lottery, which is believed to have been broadly similar to the modern casino game Keno. To play this early lottery, you would choose 5 out of 80 Chinese characters, hoping to match those drawn. The winning characters were communicated to outlying locations by messenger pigeon, which gave the game its name of *baak-gap-piu*, Cantonese for 'white dove tickets'.

In the modern variant of this lottery, Keno, the range of 80 possible numbers is preserved, but players are now permitted to select as many numbers as they choose, typically up to 15. Twenty numbers are then drawn either physically or electronically, and the player is rewarded according to how many matches (or 'catches') are made, the payoffs varying according to how many numbers were selected. Sometimes a complete miss (no matches) will also attract a small prize. The odds on Keno in casinos are poor, typically returning about 80 per cent of the stake on average, and savvy gamblers avoid the game.

At dinner parties Rome's first emperor, Augustus, would sell lottery tickets for various items and paintings to the highest bidder. When he wished to raise funds for repairs to the city of Rome, he realized that the level of taxation was so high that to raise further taxes would be politically unwise. Instead, he arranged a similar lottery with valuable objects as prizes, rather than cash, making it more akin to a raffle than a proper lottery. Then, as now, a State-run lottery was a form of voluntary taxation.

In the United States, from the late nineteenth century, before state lotteries were permitted, illicit lotteries of various kinds were popular. Early forms were known as 'policy' games and involved the selection of 12 or so numbered balls. Since these lotteries were illegal, the drawing of numbers could not be done in public, and this made it quite easy to fix the results. This led to a loss of confidence on the part of the playing public, and so a different way of selecting winning numbers was needed, one that was trusted, public, and acceptably random.[76] The result was a variety of ingenious methods: for example, taking three digits from the middle of the published number of shares traded on the New York Stock Exchange that day or the final digits from three numbers published at the end of each day's racing at the local racetrack.[77] And so this lottery became 'the numbers game'. Ticket-buyers selected a three-digit number, and so had a one in 1,000 chance of matching the winning number. Typically, winners would be paid out at 540 to 1. That made the edge against the player a whopping 46 per cent, 6 per cent of which typically went to the runner who collected the bets, the remainder to the organizer.

Notable chances

Chance of winning in a three-digit numbers game:
Probability: 0.001
Percentage: 0.1%
Rounded-off proportion: 1 in 1,000

Approximate fair betting odds: 1,000 to 1

Playing 200 days a year, over 50 years, each time with probability 1 in 1,000, on average there will be 10 wins.
Payout on a win: 540 to 1

---

[76] These qualities might remind you of the criteria for the randomness beacons discussed in the chapter 'Random Thoughts', p. 65.
[77] The numbers were the totals of the bets placed that day on Win, Place, and Show: sufficiently unpredictable to pass for random.

Margin against the player: 23 in 50 (46%)

Representative example:

10,000 chances for hits
Hits in this example: 12 = 0.120% (expected average: 10 = 0.100%)

(As before, notice that the wins are far from being evenly distributed.)

All lotteries, whether State-sanctioned or illicit, paid out far less than they took in. Lotteries are for raising funds. Since the payouts are always substantially less than the takings, each ticket is an uneconomic proposition for the buyer, when considered from an 'average winnings' point of view. Yet lotteries remain extremely popular. Large numbers of people buy tickets regularly—the possibility of a big win outweighs the near certainty of a small loss. Or perhaps the thrill of playing, of buying a chance of winning a large enough amount of money to change your life, and the anticipation and excitement of checking the result are reward enough.

What are the chances of winning the lottery?

To calculate the chances of winning a lottery is not difficult. As ever, it amounts to calculating how many ways the numbers could appear, and how many of those are successful hits. Take the UK National Lottery. There are 59 numbered balls, from which 6 are selected. If a player holds a ticket with numbers that match all six selected by the lottery machine, then that ticket entitles them to a share in the main prize. Different levels of prize are available for partial matches.

The known probability of selecting each ball means that the odds of your ticket matching the draw can be calculated arithmetically: the chance of winning each class of prize is fixed and known. What is not known in advance is the absolute amount of the main prize, the jackpot, which is distributed among those whose tickets are exact matches—and which is rolled over if there are no winners. Since there can be more than one jackpot prize winner, that main prize needs to be divided among however many winning tickets there are.

Getting technical: calculating the chances of winning the lottery

How many different possibilities are there for the draw? The first ball could be any one of the 59. The second could be any of the 58 remaining. The third ball has 57 possibilities, the fourth has 56, the fifth has 55, and the sixth has 54. Multiply all these together to reach a total of 32,441,381,280 ways. More than 32 billion. That's a big number.

How many of those ways will match your ticket? A winning ticket is a winner regardless of the order in which the balls popped out of the machine. In the televised presentation of the draw, the balls are quickly rearranged into numerical order before the final result is announced. In how many different orders could the winning numbers arrive? The first to arrive could be one of the six winning numbers, then the second could be one of five, the third one of four, the fourth one of three, the fifth one of two, and for the last there is only one ball left. Multiply those together to find how many winning ways there are: 720. So the chance of one of those ways coming up is one in (720/32,441,381,280), or around 1 in 45 million.

The chance of winning the UK National Lottery is around 1 in 45 million. The organizers estimate the typical jackpot as £5,000,000 for a stake of £2. Those look like terrible odds, but the main jackpot is not the only way of winning. Prizes are available for partial matches, and other combinations involving a so-called bonus ball. Taking all the possible ways of winning together, the lottery will pay out around 95p for every £2 ticket. That makes the margin against the player around 52.5 per cent. That's not a sensible bet to take in its own right, but then we already know that lotteries are all about raising funds for the organizer. In the case of the UK National Lottery, these funds go to good causes that would otherwise stand little chance of attracting funding from the central government.

Notable chances

The chance of winning the UK National Lottery 'Lotto' competition. Five balls drawn from 59:
Rounded-off proportion: 1 in 45 million

EuroMillions Lottery. Five balls drawn from pool of 50 plus two 'star' balls drawn from 12:
Rounded-off proportion: 1 in 140 million

US Powerball Lottery. Five balls drawn from 69, plus one powerball drawn from 26:

> Rounded-off proportion: 1 in 292 million
>
> US Megamillions Lottery. Five balls drawn from 70, plus one powerball drawn from 25:
>
> Rounded-off proportion: 1 in 303 million

In poker, a flush beats a straight

Some would argue that poker is not a gambling game at all—that it's simply a game of skill where the conventional way of keeping score is with money. Chance plays its role, but is never the deciding factor. In poker the initial deal is followed, in ways that depend on which version of poker is being played, by a number of open or hidden card draws that allow you to develop your hand. So the hands that are put on the table at the final showdown are not random, but have been built from random inputs, by players exercising their skill.[78] It's not possible to use pure probability theory to calculate the mathematical chances that if you're holding a full house you will win, in the way that we could calculate winning chances for craps or roulette. The skill elements are all-important. It's the large numbers at work again. Play enough games and the random fluctuations are smoothed out, and the underlying truth (the level of skill) has a chance of emerging.

Nonetheless, it's instructive to think about the probabilities associated with various types of poker hands. There are 2,598,960 possible five-card hands that can be dealt from a normal deck of 52 cards. Four of these hands (or 1 in 649,740) are royal flushes (A K Q J 10 in a single suit), which is the top hand in the game of poker.[79] So we can say that the probability of being dealt a royal flush at the start of a game of five-card poker is around 0.00015 per cent. At the other end of the poker scale, the number of hands with no special features, not even a pair, is 1,302,540—just over half of them— 50.12 per cent. Here are the chances of being dealt the various recognized combinations:

---

[78] And part of that skill is knowing when to fold. After the winner, the person who 'comes second', who loses least in a poker hand, is the one who folds earliest.

[79] If two players have the same hand, but in different suits, the pot is split. There is no precedence of suits in poker.

Hand	Probability	Approximate odds
Royal flush	0.00015%	650,000 to 1
Straight flush (not royal)	0.0014%	72,200 to 1
Four of a kind	0.024%	4,200 to 1
Full house	0.14%	700 to 1
Flush	0.20%	500 to 1
Straight	0.39%	250 to 1
Three of a kind	2.11%	46 to 1
Two pairs	4.75%	20 to 1
One pair	42.26%	11 to 8
Nothing (high card wins)	50.12%	1 to 1

Some of these stand out as memorable:

Notable chances

Chance of dealing five cards that form a **flush** in poker—5,108 in 2,598,960 possible hands:

Probability: 0.0020

Percentage: 0.20%

Proportion: 1 in 509

Approximate fair betting odds: 500 to 1

Imagine 200 poker players, each of whom is dealt 50 hands of 5 cards. Then, on average, of those 10,000 hands there will be 19.7 natural flushes dealt.

Here's a way to visualize the rarity of such hands. 21 black dots mark the natural flushes in this simulation of 10,000 hands.

Chance of dealing five cards that form a **straight** in poker—10,200 in 2,598,960 possible hands:

Probability: 0.0039

Percentage: 0.39%

Proportion: 3 in 764

Approximate fair betting odds: 250 to 1

On average, of 10,000 hands there will be 39.2 natural straights dealt. Here's how that might look (43 in this example):

Chance of dealing 5 cards that include **two pairs** in poker—123,552 in 2,598,960 possible hands:

Probability: 0.0475

Percentage: 4.75%

Proportion: 28 in 589

Approximate fair betting odds: 20 to 1

On average, of 10,000 hands there will be 475.4 natural two-pairs dealt. And visually (496 in this example):

> If we think of the columns here as representing 200 players, each receiving 50 hands, then 16 of them never received two pairs, while for three of them it happened seven times, and two of them had runs of three two-pairs in a row.

Playing the horses

The race is not always to the swift, nor the battle to the strong, but that's the way to bet it.

Damon Runyon

In the English Lake District, overlooking Longsleddale, Kentmere, and Haweswater, is the fell[80] known as High Street. It's said that the Romans chose it as a route through this beautiful and rough area of Britain for being a long, flat, elevated ridge. Despite its height, in the eighteenth and nineteenth centuries it was the location every year on 12 July for a fair, where locals would return to their rightful owners the sheep that had strayed from one valley to another, and where they would then compete with each other in sports such as wrestling. And they would race horses. Indeed, the rounded summit of this gently sloping peak is known as Racecourse Hill. Where there are horses, there will be racing. And where there is horse racing, there will be betting.

What are the odds?

As I write this, it is three minutes before the start of a horse race at Pontefract in West Yorkshire. The runners and their odds are these:

Horse	Quoted odds
Patrick	7 to 2
Socialites Red	9 to 2
Highly Sprung	5 to 1
Spirit Power	11 to 2
Decision Maker	15 to 2
Suitcase 'n' Taxi	9 to 1
Kibaar	10 to 1
One Boy	14 to 1
King Crimson	25 to 1

[80] Hill or mountain? You decide—its altitude is 829m above sea level.

Now if these offered odds were fair reflections of the chances of winning, with no advantage to the bookmaker or the bettor, we could interpret them as expressions of probability and calculate the implied chance of each horse winning:[81]

Horse	Quoted odds	Implied chance
Patrick	7 to 2	22.2%
Socialites Red	9 to 2	18.2%
Highly Sprung	5 to 1	16.7%
Spirit Power	11 to 2	15.4%
Decision Maker	15 to 2	11.8%
Suitcase 'n' Taxi	9 to 1	10%
Kibaar	10 to 1	9.1%
One Boy	14 to 1	6.7%
King Crimson	25 to 1	3.8%
Total		113.8%

But hold on—the total of those percentages comes to more than 100 per cent! The simple explanation for this is that the odds are *not* fair representations of probability. For a start, they include a built-in margin for the bookmaker, who is in business to provide a service to the punters and is entitled to take a cut, so the odds are less generous than they would otherwise be, implying an inflated probability of winning in each case. The excess over 100 per cent is called the overround, and is a measure of the bookmaker's cut.

But we're mistaken if we're expecting the odds to be anything like a pure reflection of the bookie's view of the probabilities of each horse winning, even allowing for the overround. If you were to keep an eye on the odds in the period before the race starts, you would see that they were constantly changing. In the event of this afternoon's race at Pontefract, the odds on Socialites Red, which had been 9 to 2, shortened to 4 to 1 in the three minutes before the start. What caused the odds to change in this way? It's not that the bookmaker had a change of opinion about the horse's prospects. Rather it was more likely a reflection of the volume of bets being placed. Betting was heavy on Socialites Red, and the bookies responded by shortening the odds, reducing the payout promised, in order to keep their books balanced.

---

[81] The formula for calculating probability when odds are expressed as A to B:

Probability = $100\% \times B / (A + B)$; so odds of 9/1 are $100\% * 1 / (9 + 1) = 10\%$.

The odds quoted on any horse are not the mathematical equivalent of the probability of that horse winning; they are simply the terms on which the bookie is offering to do business. The most that can be said is that, to the extent that the quoted prices reflect the weight of the betting, they also reflect, more or less, the balance of opinion of the betting punters (after adjusting for a margin to cover the bookmaker's earnings). If you compare your view of the nag's chances to the quoted odds, you can decide if it's worth a bet. If you reckoned that Suitcase 'n' Taxi had a better than 10 per cent chance of winning, then even though that horse was quite unlikely to win, the odds of 9 to 1 would represent objectively good terms, and therefore a rational bet. It's not just a matter of choosing the horse that is most likely to win: you need to consider the odds on offer.

In fact, Socialites Red won that race, and the starting price (the consensus odds offered by a range of bookmakers at the start of the race) ended up at 3 to 1, which made that horse the favourite when the race started. Kibaar did well to come second with a starting price of 11 to 2, and Highly Sprung came in third at 9 to 2 SP.

You don't have to place a bet with a bookie to gamble on a horse race. You can use pool betting. This is much like a lottery, whereby all the bets of a certain class are pooled together and the total pool (less an administrative fee) is divided among those holding winning betting slips. This is variously known as pool betting, pari-mutuel betting, or tote (short for totalizator) betting. In this arrangement the payout is entirely arithmetically determined. There are no quoted odds, but if a favourite wins, the dividend will be smaller, since more people will need to share the pool. If an outside chance wins, the payout will be bigger, shared among fewer winning tickets. For the 4.50 at Pontefract this afternoon, the payout for Socialites Red was £3.70, equivalent to odds of 2.7 to 1.

Combination bets

As with other forms of gambling, much of the complexity and apparent mystique of playing the horses has to do with elaborate and sometimes baffling combinations of bets.[82] You can back a horse to place, not only to win, or you can choose an 'each way' bet, which is a bet on both win and place—both tickets will pay out if your horse wins. Accumulator bets roll the proceeds of a win over

[82] How many positions count as 'places' depends on the number of horses racing. For a large field, up to four places may be recognized.

to a subsequent race. There are bets where multiple horses can be nominated, where a win by any of the listed horses will count. These complications make for ample opportunity to create very complex bets. The rules for understanding these combination bets are the same as those in the casino: combining mutually exclusive events (betting that either horse A or horse B wins in a given race) means adding probabilities and combining independent events (horse C to win in race 2, and then horse D to win in race 3) means multiplying them.

The probability of a sequence of independent events all occurring is calculated by multiplying their probabilities. For an accumulator bet, any winnings are taken forward as the stake for subsequent races, and so the winnings multiply in accordance with the odds. It was a remarkable afternoon in horse racing when in September 1996 at the Ascot racecourse, jockey Frankie Dettori rode winners in all seven races run that day. Here's how a pound bet would have accumulated for a Frankie fan:

Horse	Bet	Odds	Payout (including original stake)	Probability suggested by odds
Wall Street	£1	2 to 1	£3	33.33%
Diffident	£3	12 to 1	£39	7.69%
Mark of Esteem	£39	100 to 30	£169	23.08%
Decorated Hero	£169	7 to 1	£1,352	12.50%
Fatefully	£1,352	7 to 4	£3,718	36.36%
Lochangel	£3,718	5 to 4	£8,365.50	44.44%
Fujiyama Crest	£8,365.50	2 to 1	£25,096.50	33.33%

So, what were the chances of that happening? Based on the accumulation of bets at the prevailing odds, you might think something of the order of 1 in 20,000. But before the last race, Frankie commented, 'I am red hot'. So perhaps it would be a mistake to regard the races as independent trials in a statistical sense. Many factors contribute to a winning run, and the jockey is not the least of those. A fired-up Frankie determined to win would surely have made a big difference.[83]

---

[83] As I write this in June 2019, I hear that Frankie Dettori has today ridden four winners at Ascot, 23 years later.

Is a horse race random?

In what sense is a horse race a random event? How much of the outcome is determined by knowable information? Frivolously, you might choose to bet on a horse for their name, or the colours worn by the jockey, or your lucky number: a blind choice. But if you have a little knowledge, you might inspect the race card to check out the form of the various horses: how well the horse has performed in recent races. If you're at the race track in person, you can take a look at the horse for yourself, and make a judgement as to their condition and temperament. You can also study the patterns of the betting as a market. Just as in an investment market, flows of money can give clues to insider knowledge. A rush of money might indicate that someone who knows more than you do has taken a favourable view of the horse's chances. And if you judge that the chances are better than the probability implied by the odds offered, then it might be worth a bet.

But races are not run on paper. Even if your judgement is impeccable, random factors will come into play. A divot of turf out of place, a flash of light from a random reflection, a distracting noise. Any of these could affect the outcome, making the result unpredictable.

Political prediction markets

In the fair casino, we can do our calculations, and understand the operation of pure chance. There is little scope for subjective judgement. Prediction markets offer a very different kind of bet. If you have a strong belief that a certain public event will or will not occur, prediction markets offer an opportunity to put your money where your mouth is.

For example, at the time of writing, in 2019, on the prediction market betfair.com there is an open proposition that allows you to place a bet on whether or not President Trump will leave before the end of his first term (for whatever reason). For a bet of 'Yes', the market is offering odds of 11 to 4. For every $4 staked, an early Trump departure will secure you winnings, over and above your stake, of $11. We can easily translate this to an implied percentage: $4 / (11 + 4) =$ (approximately) 27 per cent.

Note that this is a market, and it works in the same way as other financial markets like the stock market. It balances buyers and sellers, those holding opposing views on the proposition, and the market mechanisms work to ensure that the price is set automatically at a point of equilibrium. The odds are not determined by the opinion of some political pundit, translated into probabilities. The company providing the betting platform has no stake in the betting themselves; they are simply providing the platform for willing participants to do deals among themselves. If plenty of people feel that, at the price shown, a bet of 'Yes' is underpriced (that is, Trump is *more* likely to have his term cut short, and the price should be closer to reflecting a 50/50 proposition), then they will 'buy' the underpriced proposition. Market mechanisms will then work to adjust the price until the buyers and sellers (yessers and naysayers) are once again in equilibrium.

In theory, then, the prediction markets should serve as accurate gauges of opinion, reliable measures of subjective probabilities, driven by the rational, averaged choices of people who are prepared to back their opinions with money. In practice, the volumes and stakes involved are relatively small (for the Trump bet described above, the largest bet I could place today at that price would be £20). The effect of this is that the price can easily be manipulated for a relatively modest outlay, should a political player wish to fake an impression of changing sentiment.[84]

At the same time, the online bookmakers Ladbrokes are offering odds on the inverse proposition, that Donald Trump will serve his first term in full, at 2 to 5. In this case the odds are being set explicitly by Ladbrokes, who will be counterparties to the bet. Their odds will be influenced by the weight of betting one way or another, but not necessarily in an automatic way. The odds quoted today imply a percentage chance of an early exit as 2 in 7 or roughly 28 per cent, not much different from the terms available on Betfair.

But this leaves a question to be answered. What does such a probability even mean? A frequentist argument feels nonsensical. We cannot say that in 27 per cent of cases where Donald Trump is 23 months from the end of his term, he will fail to serve those 23 months. It's hard even to say (as you might in regard to, for example, football matches) that in 27 per cent of similar cases, a Trump departure would be the outcome. Does this—the early termination of a presidency—even count as a chance event? Surely it cannot be. Political actors will act, stories and narratives will emerge, to drive towards one outcome or another. We are merely using a percentage as a measurement of our ignorance, in the face of unreckonable complexity.

---

[84] It's thought that this was done during the 2008 Presidential Election with the aim of undermining pro-Obama sentiment.

Unusual wagers

wager (n.)

'a promise, a vow, something pledged or sworn to'; also 'a bet, a wager; stakes, something laid down as a bet'. From

wajour (c. 1300). From,

wageure (Anglo-French). From

wagiere (Old North French) 'pledge, security'

The plot of Jules Verne's *Around the World in 80 Days* is driven by a wager: Phileas Fogg makes a bet with fellow members of the Reform Club that, in the light of recent developments in steamer and rail travel, it has become a realistic proposition to journey around the world in 80 days. He goes all-in: to lose this wager would be to lose all he has, and this jeopardy propels the narrative forward at breakneck speed.

Phileas Fogg is an invention, but the set-up is grounded in truth. London's clubs have long been notorious venues for wagering. White's Club, established in 1693 by Francesco Bianco ('White' is the anglicization of his surname) as a chocolate-drinking establishment, became a private club known for the eccentric behaviour of some of its members. They were particularly noted for their fondness for betting on almost any event that caught their attention, the wagers being recorded in the club's books. An example: 'April, 1819. Sir Joseph Copley bets Mr. Horace Seymour five guineas that Lord Temple has a legitimate child before Mr. Neville.' The book records one occasion when a stranger collapsed on the street outside the club's door, and bets were taken as to whether he was alive or dead. Lord Mountford and Sir John Bland made a bet as to which of two older members would die first: both of the gamblers died before either of those they were betting on. Lord Alvanley placed a bet of £3,000 on a race between two raindrops rolling down a window pane. At White's, nothing was too trivial, or too serious, for the placing of a bet.

But placing a wager need not be frivolous. After all, to risk money on a proposition indicates a degree of commitment. Putting your money where your mouth is separates genuine belief from big talk. In 2002, as part of the Long Now project, an online forum (longbets.org) was established for public, high-profile, long-term wagers, mostly concerning highly speculative questions. The *Long Bets* website allows members to make challenges, claims that can be contested by others holding opposing views. The first wager recorded on *Long Bets* was between Mitch Kapor (entrepreneur and founder of the Lotus software company) and Ray Kurtzweil (inventor and pioneer of optical character recognition). Kapor's challenge reads: 'By 2029 no computer—or "machine intelligence"—will have passed the Turing Test.'[85] At stake is $20,000, and the question is as yet unresolved. The point of the website is not gambling, but debate. Far more interesting than the bet itself are the arguments published in favour of and against the proposition. But the wagering aspect is important, for it is the wager, and the need for the terms of the wager to specify detail, that drags those arguments away from airy speculation to hard prediction.

As its name suggests, Long Bets is a forum for long-term thinking. Bet number 11 in the list is: 'At least one human alive in the year 2000 will still be alive in 2150.' That bet will not pay off for another 130 years. On the other hand, some wagers are now close to resolution. One reads: 'Driverless cars will be commercially available in Las Vegas, NV by May 27 2024. Trips may be point to point outside of the city center with no requirement for any passenger to take over manual control of the vehicle.'

But not all wagers reflect belief in a proposition: sometimes the contrary is true. In 1974 physicist Stephen Hawking bet fellow physicist (and Nobel Laureate) Kip Thorne that the X-Ray source Cygnus X-1 would turn out *not* to be a black hole. It wasn't that Hawking believed this to be true. On the contrary, he was fairly certain that Cygnus X-1 *was* a black hole. He made the wager as a kind of insurance policy, to provide some consolation if he turned out to be wrong. As things turned out, it was a black hole, and Hawking cheerfully conceded the bet in 1990. He settled up, as originally promised, by paying for a year's subscription to *Penthouse* magazine to be sent to Thorne.

Answer to:
How long have you got?

The Gambler's Fallacy is the mistaken belief that any run of losses must increase the chance of subsequent wins. It's a belief in a 'law of averages', the idea that everything must even up in the long run. It stems from a misunderstanding of the law of large numbers which holds, correctly, that given a long enough run, the *proportion* (but not the absolute number) of hits to misses will tend to converge on the underlying probability: flip the coin often enough, and the proportion of heads will get, on average and over the long run, closer and closer to 50 per cent.

[85] The Turing Test is a test of whether a computer can think. Briefly, it says that if a computer can engage in a text-based conversation and pass as human, fooling a determined interrogator, then we should regard it as genuinely intelligent.

And so it is with this game proposed by the sinister stranger. Play long enough, and the ratio of the number of times you win and the number of times he wins will converge on a ratio of 27 to 25 (10 can be rolled in 27 ways with three dice, 9 can be rolled in 25 ways). In the long run, you will win more often than he will.

But do you have the resources to play the game for the long run? You have only five pennies, but this devilish fellow hasn't disclosed to you how many he has. Your long-run reckoning is of no use if you cannot play for long enough to make your advantage count. In fact, if he has 10 pennies to play with, then even though you have the better of the odds, he is more likely to take all your pennies than you are to bankrupt him, simply because his pockets are deeper. The calculations are complex, but with 10 pennies to your 5, he has a marginal edge: a 53.35 per cent chance of winning.

His margin may be better than that. He's dropped a hint that his supply of pennies may be limitless. Recall his words: 'We play until you run out of pennies.' If indeed he can never be bankrupted, then the chance of you being ruined can be calculated to be around 68 per cent. What happens the other 32 per cent of times? You and the stranger play forever. The game never ends.

This is known as the Gambler's Ruin. The player with deeper pockets (and that might mean the casino) always has an additional advantage of being able to survive longer runs of bad luck.

CHAPTER 8

Danger of Death

Everybody going to be dead one day, just give them time.

Neil Gaiman, *Anansi Boys*

Short time to live?

The life expectancy of a baby born in the United Kingdom at the start of the twentieth century was around 47 years. For a child turning 10 in the year 1900, though, the average remaining years of life would have amounted to around 51 years, giving a forecast average age at death of 61. How could it be that in 1900 a 10 year old could look forward to more years of life ahead of them than a newborn?

On average, how many years would you think remained for a young adult, turning 20 years old in that year?

How many more years might be in store for an older person, turning 50 in 1900, who had already outlived the life expectancy of a baby?

Answers at the end of this chapter.

Certain but uncertain

Many people have said it, but Benjamin Franklin is probably the true source of the aphorism 'in this world nothing can be said to be certain, except death and taxes'.

What are the Chances of That? How to think about uncertainty. Andrew C. A. Elliott,
Oxford University Press. © Andrew C. A. Elliott 2021. DOI: 10.1093/oso/9780198869023.003.0008

In the long run, of course, Franklin was right, about death at least. Nobody lives forever. But it is far from certain *how* people die, and *when* they die. Naturally, these questions are a source of worry. The days of our lives are a precious and limited resource, always at risk of being cut short. But death is not entirely a random process, and understanding how and when people die can be invaluable, not only for ourselves, but also for public policy, to know where medical and public health efforts are best directed.

What did we die of?

John Graunt, whom we met earlier, was one of the first people to study population change. He developed an intense interest in the statistics of births and deaths in seventeenth-century London, and this culminated in the publication of a volume entitled *Natural and Political Observations Made upon the Bills of Mortality*. The Bills referred to were weekly returns prepared by parish clerks for the church parishes in and around London, and summarized annually. He noted the popular interest in these, but could see that there was more to them than the superficial interest that most people took, and that they might yield interesting conclusions if analysed systematically. Graunt was a haberdasher, a merchant in small items for sewing, and he took a keen interest in his city, London, and the living conditions there. He saw that an understanding of mortality and the causes of death was important to the health of the city, and he understood that his analysis could be of value.

A good data analysis must always start with understanding the source of the data. Graunt records the method of collection in his *Observations*:

> When any one dies, then, either by tolling, or ringing of a Bell, or by bespeaking of a Grave of the Sexton, the same is known to the Searchers, corresponding with the said Sexton. The Searchers hereupon (who are antient Matrons, sworn to their office) repair to the place, where the dead Corps lies, and by view of the same, and by other enquiries, they examine by what Disease, or Casualty the Corps died. Hereupon they make their Report to the Parish-Clerk, and he, every Tuesday night, carries in an Accompt of all the Burials, and Christnings, hapning that Week, to the Clerk of the Hall. On Wednesday the general Accompt is made up, and Printed, and on Thursdays published, and dispersed to the several Families, who will pay four shillings per Annum for them.

He provides a considered discussion of the reliability of the Searchers who were the primary source of this information. In some cases he is pragmatic about inconsistency, commenting:

In case a man of seventy five years old died of a Cough (of which had he been free, he might have possibly lived to ninety) I esteem it little errour (as to many of our purposes) if this Person be, in the Table of Casualties, reckoned among the Aged, and not placed under the Title of Coughs.

In other places he agonizes over definitions and interpretations of symptoms, as here, where he questions the proper classification by unreliable Searchers of those who have died from 'Consumption' (tuberculosis):

all dying thereof die so emaciated and lean (their Ulcers disappearing upon Death) that the Old-women Searchers after the mist of a Cup of Ale, and the bribe of a two-groat fee, instead of one, given them, cannot tell whether this emaciation, or leanness were from a Phthisis, or from an Hectick Fever, Atrophy, &c. or from an Infection of the Spermatick parts.

Graunt shrewdly drew a distinction between 'Epidemical' diseases (such as the plague) and 'Chronical' diseases (such as tuberculosis), and commented:

We ventured to make a Standard of the healthfulness of the Air from the proportion of Acute and Epidemical diseases, and of the wholesomeness of the Food from that of the Chronical.

The rate of epidemical deaths, as the name suggests, varied substantially from week to week, whereas the rate of chronical deaths in proportion to the population remained relatively constant. Graunt had worked out the difference between the acute and epidemical diseases that came and went, and which he concluded were due to something in the air, and the chronical diseases that persisted at more or less the same level, and which were due to some stable environmental condition (he thought it might be 'the wholesomeness of the Food'). Working before the germ theory of disease was understood, he had discerned the difference between infectious causes and environmental causes. His work shows how a statistical analysis can give clues as to the cause and mode of transmission of disease.

For the first edition of his publication, in 1662, John Graunt selected a series of years (1629–1660, excluding 1637–1646[86]) and performed a consolidated analysis of cause of death. He provides a variety of summaries, and commentary on them. Here is a summary of his summaries:

---

[86] He explains the missing years thus: 'MEMORANDUM, that the 10 years between 1636 and 1647 are omitted as containing nothing extraordinary'. Not an approach that would be considered sound nowadays!

Cause by category	Number	Percentage of total
Diseases primarily affecting children	71,124	31.0%
Small Pox, Swine Pox, Measles, Worms without Convulsion (estimating that half of these would be children under the age of 6)	12,210	5.3%
Acute diseases, excepting the Plague	About 50,000	21.8%
Plague	16,384	7.1%
Chronical diseases	About 70,000	30.5%
Notorious diseases	5,550	2.4%
Casualties (accidental and violent deaths)	2,889	1.3%
All deaths (Graunt's total)	About 230,000	100.0%

In regard to those last two categories, 'Notorious' diseases and 'Casualties', he comments:

> whereas many persons live in great fear and apprehension of some of the more formidable and notorious diseases following; I shall only set down how many died of each: that the respective numbers, being compared with the total 229,250, those persons may the better understand the hazard they are in.

Graunt is debunking exaggerated fears: showing that the diseases that cause the greatest anxiety (including Apoplexy—1,306 deaths in 22 years, and Leprosy—6 deaths) are relatively few in number. Violent deaths were also remarkably few. There were 829 Drowned, but only 86 Murdered. Just 51 Starved, and 7 were Shot[87] in the 22-year period he covered.

Plague and fire

When John Graunt published his tables of deaths in 1662, he could not have expected that within five years, the pattern of deaths in London would be dramatically disrupted. For example, consider the number of deaths in this table based upon the Bills for 1662 to 1669:

---

[87] 'Shot' is noted separately from Murdered. Possibly these were accidental shootings?

Area of London	1662	1663	1664	1665	1666	1667	1668	1669
Within the walls	3,123	3,002	3,448	15,207	1,977	761	796	1,323
Without the walls	6,104	5,608	7,168	41,351	4,543	6,428	6,865	7,501
Middlesex and Surrey	4,689	4,269	4,885	28,554	4,465	5,610	6,521	7,219
City & Liberties of Westminster	2,638	2,477	2,796	12,194	1,753	3,043	3,096	3,223
In Total	16,554	15,356	18,297	**97,306**	12,738	15,842	17,278	19,432[a]

[a] Includes 166 buried in 'demolished parishes'

It doesn't take much of a statistician's eye to see, directly from these numbers, that something horrific happened in London in 1665. The number of deaths rose, in all areas of the city, to something like five times the level for previous years. We'd call this an excess mortality rate of around 480 per cent. The cause of this increase? 1665 was the year of the Great Plague, which hit London and its surroundings hard. John Graunt estimated that the population of Greater London, before the plague struck, was around 380,000 people, making the recorded death rate for 1665 around 25 per cent. Moreover, he notes that the recorded figures for deaths are likely to have been considerably underestimated. It is difficult to imagine the disruption that the disease must have caused to the life and business of the city. Here is the analysis of cause of death, as published in the Bills of Mortality for 1665, the year of the plague:

In that year, the plague accounted for 69,000 of the total of 97,000 deaths (71 per cent). Here are the top causes of death in each of the years 1662 to 1669:[88]

Cause	1662	1663	1664	1665	1666	1667	1668	1669
Plague	12	9	6	68,596	1,998	35	14	3
Consumption and Tissick	3,485	3,260	3,645	4,808	2,592	3,087	2,856	3,162
Ague and Fever	2,601	2,107	2,258	5,257	741	947	1,268	1,499
Griping in the Guts	835	866	1,146	1,288	676	2,108	2,415	4,385
Convulsion and Mother	1,053	1,011	1,181	2,036	825	1,210	1,420	1,732
Teeth and Worms	1,032	979	1,122	2,614	715	914	1,178	1,463
Chrisomes and Infants	1,315	1,299	1,491	1,258	749	901	753	755
Aged	1,215	1,171	1,154	1,545	611	952	863	964
Dropsy and Tympany	990	934	1,079	1,478	983	1,134	948	871
Flox and Small Pox	768	411	1,233	655	38	1,196	1,987	951
Surfeit	186	191	244	1,251	345	417	438	446
Spotted Fever and Purples	148	128	116	1,929	141	96	148	114

There is much to wonder at in this table, not least the unfamiliar names of the diseases. The top cause of death over these eight years is the plague, and it is notable just how insignificant a factor it was outside the two worst years. There is also an increase in the plague year in deaths from 'Surfeit', and also 'Spotted Fever and Purples'. Could some of these have been mis-recorded plague cases? And why was there an increase in 'Griping in the Guts' in 1669?

It is perhaps not quite as easy to see from the figures the traces of what happened in London the next year, 1666, the year of the Great Fire. Astonishingly, in a fire in which 70,000 people are thought to have lost their homes, only 6 are officially recorded as having died. Where the plague claimed lives, the fire destroyed property, and was confined to the area 'within the walls'. But looking at the figures for deaths within the walls in 1667 and 1668, the years after the fire, they are at levels around a quarter of what they had been before the plague struck (whereas deaths 'without the walls' are comparable to pre-plague years). If the deaths within the walls were a quarter of what they had been, it is likely

---

[88] I've included any cause of death that exceeded 1,000 deaths in any one year.

that this reflects a similar fall in the size of the population within the walls. What a picture these numbers paint of a burned out city, its diminished population starting the task of rebuilding!

The wording of the summary of the Bills of Mortality for 1666 reveals the impact of the Great Fire. The statistics for christenings in that year are given not as a single number, but as two figures, noted against separate headings, covering periods before and after the fire.

> Christened in the ninety-seven Parishes within the Walls of London, from the 19th of December 1665, to the 28th of August 1666: **1165**

and then:

> Since which Time the late dreadful Fire hath consumed and laid waste eighty-one parishes; So that there hath been Christened in the sixteen Parishes now standing, from the 28th of August 1666, to the 18th of Decem. following: **151**

The Bills list 16 parishes still standing in 1667 and 1668, but by 1669, once again we see figures for 97 parishes, and the number of christenings and deaths in those parishes growing.

> It may be now asked, to what purpose tends all this laborious buzzling, and groping?
>
> Graunt's *Observations (opening of final section)*

Graunt's analysis of the Bills of Mortality was the work of a data scientist. He is recognized as the first demographer and the first epidemiologist, but he was undoubtedly also a pioneer in data analysis. He saw the value hidden in a meticulously collected data set, understood the need for consistency in definition, and was able to extract a series of concise and substantiated observations. His attitude is always rational. He debunks superstitions (for example, that plague years are associated with the start of new monarchs' reigns). He dampens fears of awful diseases and violent deaths by showing how rare they are, using numbers as evidence. Here is an acute mind at work, an analyst. He answered his own question ('to what purpose?') in this way:

> I conclude, That a clear knowledge of all these particulars, and many more, whereat I have shot but at rovers, is necessary in order to good, certain, and easie Government, and even to balance Parties, and factions both in Church and State.

That was then, this is now

Graunt saw the importance of statistical work in guiding public policy. That the numbers can help us in assigning priorities is no less true now than in his times. How do today's figures look? What do we die of now? For England and Wales, the Office for National Statistics does what John Graunt used to do, assembling data sets from records of deaths. For the five years 2013–2017 inclusive, these were the top 10 categories of causes of deaths.

	Cause	Number	Percentage
1	Neoplasms (cancers)	739,190	28.47%
2	Diseases of the circulatory system (including strokes)	682,035	26.27%
3	Diseases of the respiratory system	362,031	13.94%
4	Mental, behavioural, and neurodevelopmental disorders	226,638	8.73%
5	Diseases of the nervous system	138,689	5.34%
6	Diseases of the digestive system	124,445	4.79%
7	Other medical[89]	58,330	2.25%
8	Diseases of the genitourinary system	46,059	1.77%
9	Endocrine, nutritional, and metabolic diseases	37,960	1.46%
10	Certain infectious and parasitic diseases	27,353	1.05%
	All other causes	153,440	5.91%
	Total	259,6170	100%

Increasingly, we die from non-communicable diseases.[90] Not from infections (tenth in the table, and barely 1 per cent of deaths) and not from accidents or violence.[91] Our bodies fall prey to cancers; our hearts give out. By and large, these are causes associated with age. The majority of deaths, in these days, and especially in privileged societies, occur when life has been lived, when the body is worn out. Chance will of course dictate the timing and circumstance of death

[89] The classification used is the ICD-10, the tenth revision of the International Statistical Classification of Diseases and Related Health Problems. This row in the table represents 'Symptoms, signs and abnormal clinical and laboratory findings, not elsewhere classified'.

[90] I write this in the midst of the most serious infectious pandemic in a century. Certainly, the figures for deaths in 2020 and 2021 will not reflect the proportions shown in this table. Covid-19 has already taken a place high in that list. Still, this does not invalidate the general trend that progressively, we are conquering infectious diseases. We will surely learn much about how to do this from the current pandemic.

[91] The biggest non-medical cause of death is Falls: it would be 11th in the table.

even in these cases, but when death comes at what is clearly the end of a long life, it is hard to see this as an unlucky blow of fate. By and large, the avoidable causes of death are indeed avoided.[92]

When do we die?

The days of our years are threescore years and ten; and if by reason of strength they be fourscore years, yet is their strength labour and sorrow; for it is soon cut off, and we fly away.

Psalms 90:10

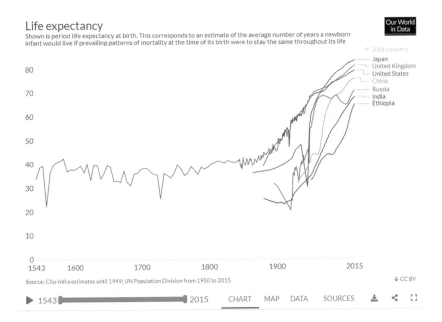

Life expectancy
Shown is period life expectancy at birth. This corresponds to an estimate of the average number of years a newborn infant would live if prevailing patterns of mortality at the time of its birth were to stay the same throughout its life

Our World in Data

+ Add country
Japan
United Kingdom
United States
China
Russia
India
Ethiopia

Source: Clio-Infra estimates until 1949; UN Population Division from 1950 to 2015

⇕ CC BY

▶ 1543   2015   CHART   MAP   DATA   SOURCES   ⬇   ⟨   ⌞⌝

For centuries, in the United Kingdom, the average age at death was in the thirties and forties. At the time that John Graunt was writing, the average age at death in the United Kingdom was around 35.[93] This does not mean that most

---

[92] Different countries have responded differently to the Covid-19 pandemic. Some have been remarkably effective at taking measures to protect their populations. In other countries, sadly, the response has been less well managed.

[93] This is generally termed 'life expectancy', but don't be misled by the word 'expect' in that term. It should *not* be interpreted as 'how long you might expect to live'. 'Expectancy' in this context should be taken as meaning something like 'average'.

people *typically* lived to an age somewhere around 35. Rather, you should think of people dying at all ages, with a broad spread. Many died young (Graunt reckons about a third), and some died old (Graunt has it that 7 per cent made it beyond 70).

One of the misconceptions that John Graunt debunked was a belief of the time, that the chance of a person surviving 10 years from their current age was a 50/50 proposition, regardless of how old they were. This equates roughly to a constant 6.7 per cent chance of dying each year. We can easily do a Monte Carlo style simulation of what that might look like. Imagine 200 people aged 20, and see what happens when each person has a 6.7 per cent chance of dying in any year. Run the simulation for 50 years and we get something like this:

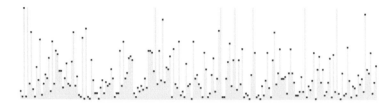

The bottom of the chart marks age 20, the top, age 70. It shows a simulation of 200 people starting from age 20 (we'll discuss childhood mortality a little later). The black cells mark deaths, and the grey vertical bars extending upwards represent years survived. This is the pattern we would see if the risk of death were entirely unrelated to age, and it does not match what we would expect from the population in Graunt's London. For one thing, the age profile shown is far too young. This picture shows too many people who die in their twenties and thirties. In reality a 20 year old would have a good chance of living a further 30 years, to give an average age at death (for those who had survived to 20) of over 50. In the chart above, we should expect the black dots to fall predominantly in the upper half of the diagram (that is, after age 45).

Let's look at the picture if we assume a more realistic pattern of mortality: one that starts low and increases with age (these are still 20 year olds, we will come to childhood deaths soon).

The difference is clearly visible. In this case, we start with a risk of death for a 20 year old of just 2 per 1,000 per year, increasing by 10 per cent each year over the course of five decades, so that by the time we reach age 70, the mortality rate has reached 23 per cent per annum. This is a much better portrayal of the pattern of survival and death from age 20 that you might have seen in Graunt's London. It's clear that a low life expectancy does not mean a population where no one lives to an old age.

Let's see how things look nowadays. Here is the same sort of chart, but using parameters closer to what we would see in the United Kingdom today (the bottom of the chart still represents age 20, but I've added 20 extra years at the top to take it to age 90):

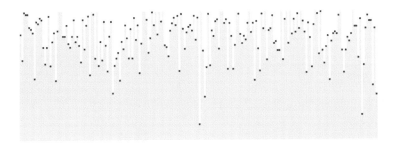

Modern medicine and public health measures mean that compared to earlier ages, only a few people nowadays die early. The mostly solid areas of grey represent the great majority of people who live into and through middle age, before death starts to take its toll in old age.

Survival

The charts presented show in a very direct way the survival patterns of 200 simulated 20 year olds: grey bars extend upwards and end in black. The chart below shows a different representation of changing mortality: survival curves.[94] Each curve relates to a group of people born in the same year and shows what percentage of these people are expected to still be living at different ages. You can easily see the ages at which most deaths happen. The lowest curve shows the pattern for people born in 1851, and you can see that in that year around

---

[94] The chart here comes from 'Our World in Data', a highly recommended source of wonderfully put-together statistics. The base data is from the United Kingdom's Office for National Statistics.

only 65 per cent would make it to age 20. Nonetheless, even then, a few people would reach age 90.

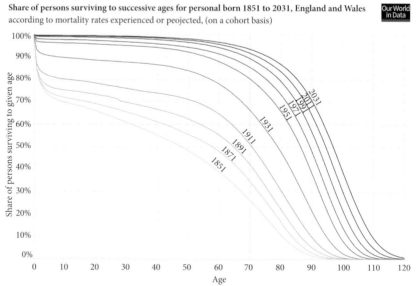

Share of persons surviving to successive ages for personal born 1851 to 2031, England and Wales according to mortality rates experienced or peojected, (on a cohort basis)

Our World in Data

Data source: Office for National Statistics (ONS). Note: Life expectancy figures are not available for the UK before 1951: for long historic trends England and Wales data are used
The interactive data visualization is available at OurWorldinData.org. There you find the raw data more visualizations on this topic. Licensed under CC-BY-SA by the author Max Roser.

Looking at the changes over the decades, the successive curves move upwards (fewer people die early) and outwards to the right (more people reach older ages). The survival rate increases: we see a much flatter curve through most of life, with far fewer deaths at early ages, followed by a steeper dropoff at older ages. In 1851, 50 per cent of people (live births) would have died by around age 47, but for those born in 2011, we expect 50 per cent to live to as old as around 94, a literal doubling of the years of life compared with those born in 1851.[95]

Childhood mortality

The survival chart is especially striking in the way it shows the improvement in childhood mortality. Survival to age 10 has risen from around 70 per cent in 1851 to close to 100 per cent now. The Bills of Mortality from which Graunt worked did not include age at death, which limited his ability to perform any

---

[95] The age by which half of the group has died is, in statistical terms, the *median* age at death.

proper analysis of how mortality progressed with age. He did, though, form an estimate of child mortality by looking at which diseases were entirely or substantially associated with death in childhood (he used age 6 as a cut-off), and he reached the conclusion that 36 per cent of live births would not survive beyond the age of 6.

Let's return to our visualization of age at death. Starting the chart at age 0 rather than 20, and incorporating a representation of child mortality into our simulation, enables us to visualize the pattern of deaths in Graunt's time. The paler cells clustered at the bottom are the childhood deaths. The chart runs from age 0 at the bottom to 70 at the top.

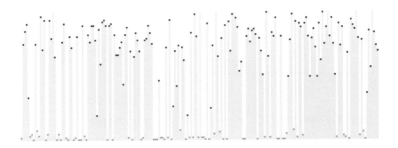

In an equivalent chart for modern conditions in the United Kingdom, the childhood deaths might not be visible at all: at a rate of ½ a per cent, on average, just one childhood death in 200 births might show as a single dot near the bottom of a chart like this. Vaccination programmes mean that childhood diseases such as mumps and measles, which were common until very recently, are now rare. Smallpox has been eradicated. Tragically, anti-vaccination campaigns have undermined confidence in these programmes, and the number of deaths from measles is once again increasing. Of all the triumphs of modern medicine and public health, the reduction in child mortality must rank as one of the most impressive, and the most meaningful.[96]

Increasingly, in privileged societies, we are managing to avoid the deaths that are avoidable. In comparison to the past, we have little reason to fear early death. And this does not only apply to wealthy nations. All over the world, rates of death in childhood have fallen dramatically.

---

[96] In passing, it is not difficult to see that the vaccination arguments are rooted in different perspectives arising from the **individual–collective** duality.

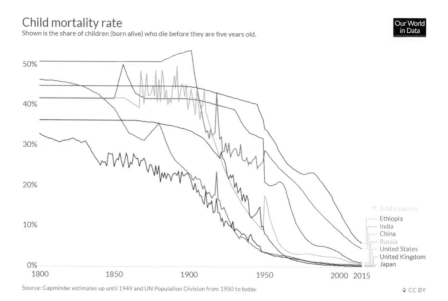

Child mortality rate
Shown is the share of children (born alive) who die before they are five years old.

Our World in Data

Ethiopia
India
China
Russia
United States
United Kingdom
Japan

Source: Gapminder estimates up until 1949 and UN Population Division from 1950 to today   ⇌ CC BY

This shows the startling drop in child mortality over the past two centuries in different parts of the world. Graunt's 36 per cent is thankfully a thing of the past, and that fact is due in no small part to Graunt himself, who started the systematic and methodical analysis of how people died.

Answer to:
Short time to live?

Why might a child aged 10 in 1900 expect to live more years than a newborn baby? Simply put, because the 10 year old was already a survivor, and had lived through what would most likely have been the riskiest part of their life: infancy and early childhood.

Consider that 10-year-old child, who had been born in 1890. Of those born in the same year, about one-quarter would not have survived to their tenth birthday. So, when calculating the average length of life for the 1890 group, that quarter who died before age 10 would contribute very few years of life to the total used to arrive at an average. Most of the years of life in that total would come from those who survived childhood, and had the prospect of longer lives ahead of them.

This chart from *Our World In Data* not only shows how life expectancy in England and Wales has changed over the years, but also shows the impact of child mortality: the large gaps between the red, orange, and yellow curves show what a difference mere survival, especially survival to age 5, makes.

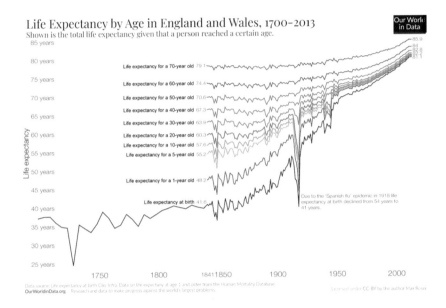

Life Expectancy by Age in England and Wales, 1700–2013
Shown is the total life expectancy given that a person reached a certain age.

Reading off the chart, you can see that a 20 year old in 1900 would expect to live to around 62 years of age, a further 42 years, while a 50 year old might be looking at a further 19 years.

The fourth duality: Uniformity–Variability

We understand averages, but we have a weaker grasp on the idea of variation. In order to understand the world, we simplify it. And so we classify and take averages and compare group with group, often making a tacit assumption that the groups are uniform. The average daytime high temperature for Surrey, in England, in June, is around 20 °C. But the days this month have varied from the frankly miserable (around 14 °C) to the sweltering (30 °C). Engineers may design a bridge to survive a 50-year flood, but not a 100-year flood. They need to understand extremes (and how they may be changing) as well as averages.

In some ways, this is a refinement of the **individual–collective** duality. It's about how you regard the individual within the collective. Do you focus on the average or the exceptional? Do you see uniform sets to which you can apply consistent rules, or do you look for the variability, and seek to accommodate that variation?

Policy determined based on national averages may not always suit the circumstances of smaller regions: this has been highlighted by the responses to the Covid-19 pandemic. It is hard not to feel that you are being unfairly treated when your local area is subjected to stringent lockdown rules simply because of severe infection rates in a city some distance away.

If this year's crime statistics are higher than last year's, the newspapers report a surge in crime, even if the change is within the typical range of random variation. But headlines need simple narratives, so the sub-editor interprets any difference as signalling a trend. Then, if next year's numbers show a fall, they'll be able to use the opposite narrative. When you read the article under such a headline, you might stop to ask: but what were the chances of that change?

Averages can be deceptive. Typically, income and wealth have skew distributions, and that makes the statistical mean a misleading number. Average household income is not the income of a typical family. To understand the numbers better, it helps to know more about their distribution: how wide a spread of values should be expected, or the extent to which it is skewed. It's worthwhile to sometimes just look at, and play with, the data a little, to get a feel for the variability.

Still, averages are easy to calculate and are often the best measure available. Sometimes an average, assuming uniformity, may serve perfectly well. But we shouldn't forget that the average hides complexity and diversity. If your analysis assumes uniformity, you may be taken by surprise when you come across extreme cases. The unexpected may be much more common than you'd expect.

CHAPTER 9

What's There to Worry About?

I am an old man and have known a great many troubles, but most of them have never happened.

Mark Twain

How long can your luck last now?

Oddly enough, that stranger at the bar is becoming a familiar figure. He's not what you'd call a friend, you don't even know his name, but it's got to the point where you feel comfortable taking your place alongside him and ordering your drink.

'Remember the other day,' he begins, 'when I placed a single joker into a shuffled deck, and we counted how many cards you could turn up before the joker showed his face?'

'Yes, I remember. The rule was that I had to flip 26 cards to win. But then you went and left before we finished.'

'Never mind that, we settled up for that. No, tonight I have a new teaser, a variation on that game for you.'

'Okay.'

'This time I'm going to shuffle not one, but two jokers into the deck. I'm going to increase the risk, and we'll see how long you can survive before either of them turns up.'

'Well, the odds are much worse for me like that. For every card that is dealt I'll have twice the risk of getting a joker. I need a much smaller target than 26 cards.'

What are the Chances of That? How to think about uncertainty. Andrew C. A. Elliott,
Oxford University Press. © Andrew C. A. Elliott 2021. DOI: 10.1093/oso/9780198869023.003.0009

'Yes, that's reasonable. Okay, what do you think the target should be, to make the game fair?'

Well? What would a fair target be?

Panic over pandemic?

A moment in time

The section that follows was written in late February 2020. At that time, the significance of the Covid-19 outbreak was starting to become apparent. It felt wrong to ignore this event and the uncertainty it was causing. What follows is a snapshot of a moment in time. It is very close to my first draft, and has been edited only for correction of language. No doubt, in the months and years to come, much will be written about this pandemic. For the most part, those writers will have the benefit of hindsight. For me, I didn't know how history would record this period—and still don't—but it seems useful to capture a moment of uncertainty, as it was and as it felt to me at the time.

We may no longer fear the Black Death, but scary diseases are by no means a thing of the past. As I write this in February 2020, an epidemic of a coronavirus strain, which has recently been given the name Covid-19, is on the cusp of turning into a pandemic. The outbreak started in Wuhan in China, but the point has now passed where there are more new cases each day outside mainland China than within. South Korea, the cruise ship *Diamond Princess*, Italy, Iran, and Japan (in that order) are the locations currently reporting the largest number of cases. I don't know how this will turn out, and that makes it an interesting opportunity for self-reflection. By the time you read this, it may have all blown over, or it may have become an episode of historic significance that has changed the world. You, my reader, have the benefit of that hindsight: I have the dubious pleasure of having an opportunity to honestly document my anxieties in the moment. For there is no doubt that I am anxious. I don't seriously fear for my life (although in the back of my mind I wonder if perhaps I should), but I am very unsettled. In particular, my personal worries are mainly about disruption to travel plans: my family and I have tickets to fly to Naples for a week's holiday in three weeks' time. It seems very possible that this trip will have to be cancelled.

I also worry greatly when I read about a 'realistic worst case scenario' that might involve hundreds of thousands, but probably not millions, of deaths in the United Kingdom. I am dismayed about the predictable reactions I already see on social media, ranging from outright dismissal of the risk ('no worse than the flu') to the hysterical ('it will wipe us all out'). Responsible news outlets do their best to summarize and encapsulate the risk information, but the truth right now is that no one knows enough. The articles that were written three weeks ago read like very old news indeed. The facts of this outbreak are still to emerge.

It seems right that the key to a rational approach is to try to keep things in perspective. And that means having something of a feel for the probabilities involved. But at this point, very little is known. We have a variety of estimates of the reproductive number R_0, ranging from 1.4 to 4.0. This is a measure of how many people, on average, will catch the virus from each infected person. We have estimates of the fatality ratio ranging from 1 per cent to 3 per cent, although at this point the ratio of deaths to recoveries (which is measuring something different) is around 1 to 12, and falling.[97] Epidemiologists are working to get the measure of this thing, to build a useful model of the way it spreads, and to calibrate the model by finding the right parameters. If they can model the process with any accuracy, then they can begin to put in place measures that might be effective in controlling the spread of the disease, or at least be helpful in planning the response if, as may be the case, a pandemic is inevitable.

Through this book, I've proposed five dualities that make us uncomfortable when we think about uncertainty. How do they relate to this highly uncertain situation?

Individual–Collective

To make a rational judgement, I am trying to seek out authoritative collective statistics, but my own anxieties revolve around my specific individual concerns. Will I have to cancel travel plans to Italy? Should my worries go beyond the annoyance of rearranging my trip?

---

[97] It's a struggle to find a meaningful definition of some of these measures. Deaths as a proportion of reported cases is not a useful measure at this stage: we must allow for the time lag involved. Then, too, in the early days, many deaths may have been due to a lack of preparedness. Epidemiologists also stress that many people may have the virus, but never show bad enough symptoms for them to report or be tested. By rights, they should be counted among those exposed to risk of death from Covid-19, but they won't be.

Randomness–Meaning

Where does this disease come from? Can we understand the patterns of who catches the disease and who escapes? Epidemiologists are frantically trying to understand how the disease spreads, what kinds of people are most susceptible, and what preventive measures might be effective. While some are spinning conspiracy theories, scientists are trying to tease out meaning from randomness.

Foresight–Hindsight

As I write this, I am painfully aware of the difficulties of prediction. In six days or six weeks or six months, much of my current anxiety over the uncertainty will be irrelevant: things will have moved on. By then, hindsight may be able teach us something new about this disease. For now, though, the impossibility of foresight is painful. I want a prediction I can have some confidence in, but for now there is nothing but uncertainty.

Uniformity–Variability

Numbers are still very small, but we are starting to understand that this disease does not have the same effect on everyone. Some experience few or no symptoms, some die. The elderly seem more affected than the young, and those with pre-existing health problems appear to be more vulnerable and suffer more.

Disruption–Opportunity

It's hard to see many opportunities for creativity arising from this devastating chance event. But hard times sometimes stimulate scientific ingenuity. Perhaps the best that can be hoped for is that this threat will spur on scientific development and leave us with new knowledge and new techniques for more effective responses to new diseases in the future. We may learn lessons.

Since I wrote those words, much has changed, in particular the nature of the uncertainty we face. But I'll resist the urge to revise and will leave those words to capture that moment in time.

Calculating the risks

The threat of pandemic is terrifying. But every day our watchful minds are filled with worries: some mundane, some existential. Potential dangers lie in wait for us every day. The knives in the kitchen drawer are sharp; travel has its risks; storms rattle through the country leaving floods in their wake; people are shot for what seems like no reason; suicide bombers take innocent lives. When a health risk is developing day by day, trying to put numbers to that risk is very difficult, but for some of these other dangers, we already have reasonably good numbers, and with a little thought, we can do the calculations needed to bring some perspective to the things we worry about.

risk (n.)

> 'possibility of loss or injury'. From
>
> *risque* (French , 16c.). From
>
> > *risco, riscio, rischio* (Italian). From
> >
> > > *riscare* 'run into danger'.

Intriguingly, this etymology lacks any sense of randomness.

Micromorts and microlives

Probabilities are often very small numbers, and we are not very good at grasping very small (or very big) numbers. How does the chance of dying in a skiing accident compare to hang gliding? What is the risk of dying while running a single marathon? Answer: around 7 *micromorts*. A micromort is a one-in-a-million chance of dying. Micromorts are a way of talking about relative risks in a consistent way with numbers of manageable size. They're the invention of Ronald A. Howard, and play off the idea of a microprobability, a one-in-a-million chance of anything.

Living for one day in the United Kingdom, or the United States, your baseline risk of death is of the order of 20 micromorts—a 20 in a million chance of dying every day. So running a marathon on a given day increases your risk from 20 to 27. It's quite crude, but it is a neat way of putting things in perspective and giving a sense of scale to the risk. Registered scuba divers expose themselves to around 5 micromorts with every dive, while an ascent of Everest will cost you around 40,000 micromorts.

In a similar way, a *microlife* is a simple way of communicating the impact of risky activities on life expectancy. A microlife is a millionth of a typical life expectancy, which comes to roughly 30 minutes. So, smoking a pack of cigarettes will cost you around 10 microlives, shortening (on average) your life by about five hours. Eating five portions of fruit and vegetables (not per day, just five portions) will boost your lifespan (on average), by 3–4 microlives. Professor David Spiegelhalter describes this kind of measure as 'rough but fair', giving a useful way of putting risks into perspective.

Poisons and potions

Observe due measure; moderation is best in all things.

Hesiod (c.700 BCE)

Whether we recognize it as such or not, whether we mean to or not, we all regularly dose our bodies with chemicals. Every bite of food we eat contains chemicals, and this is far from being a bad thing. The sugars in fruits, the electrolytes in vegetables, the oils and fats in meats and in pulses—all are chemicals and they are all processed chemically by our digestive systems and make their way into our bodies to play a role, for health or for harm, in our metabolism. There is no sharp dividing line between a poison and a potion. Too much sodium in your food will kill you; so will too little. Digitalis, derived from the foxglove plant, has long been known (and used) as a deadly poison. It is also used in heart medication. Botulinum is a deadly neurotoxin; it is also used, under the name Botox, as a treatment for wrinkles.

Broadly speaking, any substance that we put in our bodies can help us or harm us. It's all in the dose, and balancing good and bad effects is a matter of understanding the risks involved. Popular media provides us with plenty of supposedly helpful health advice about what we should and should not be consuming, and in what quantities. Often, though, the effect of this is to make us worry more: it's hard to know how much attention to pay to this guidance. Perhaps Hesiod was right—maybe moderation is best: there are few foods that do serious harm, if taken in modest amounts.

A little etymology tells us that the ancient Greeks knew about the dual nature of medication.

pharmaceutical (adj.)

'relating to medicinal drugs, or their preparation, use, or sale. From

pharmaceuticus (Late Latin) 'of drugs'. From

pharmakeutikos (Greek), from *pharmakeus* 'preparer of drugs, poisoner'

Botulinum requires a dose of 2 nanograms per kilogram to be lethal when injected—for a person weighing 100 kilograms that is just one five-millionth of a gram. Cyanide, on the other hand, requires around a fifth of a gram for a lethal dose. But there is one substance that humans absorbed in great quantities, over long periods, and that in the twentieth century killed more people than any war. That poison was cigarette smoke.

Smoking kills

Cigarettes are a classy way to commit suicide.

Kurt Vonnegut

I'm eighty-three and I've been smoking since I was eleven. I'm suing the cigarette company because it promised to kill me and it hasn't.

also Kurt Vonnegut

There is now no doubt that cigarette smoking is an extremely dangerous thing to do. Around half of all smokers die from causes related to their smoking. In the United States, around 1 in 5 of all deaths can be attributed in some way to smoking, even though smokers now make up less than 1 in 7 of the population. It was lung cancer that first alerted medics to the pernicious effects of smoking, but it's far from being the only route through which smoking kills. Heart attacks, strokes, bronchitis, emphysema, and other cancers are all strongly associated with cigarette smoking.

But there's a paradox here too, and it's related to the **foresight–hindsight** duality. Looking forward, it's not certain that if you smoke, you will die of a smoking-related condition. Smokers die of other causes too, and we all know people who seem immune, who reach a grand old age even though they are lifelong smokers. But looking backwards, if someone dies of lung cancer, then it is very likely indeed that the cause was smoking.

Here are some ways of measuring the effects of smoking on mortality:
- What proportion of all deaths are smoking-related?
  - In the USA, around 1 in 5 of all deaths
  - In the UK, around 1 in 6 of all deaths
- What proportion of deaths among smokers are smoking-related?
  - In the USA, around 50%
  - In the UK, the proportion is similar
- How much does smoking affect your chance of dying?
  - Through life, between +50% and +100% chance of dying in any year
- How much does this affect your lifespan?
  - Around 10 years of life lost, on average

It's now clear that stopping smoking always has a beneficial result. Stopping early in life is almost entirely effective. Someone who stops smoking by around 30 has the same risk profile as someone who has never smoked. Even stopping after the age of 60 will substantially reduce the harm done by smoking. Smoking is a lifestyle choice with significant risk; if you are a smoker, that should worry you. You should be worried enough to stop.

You are what you eat?

Through the twentieth century, medical advances reduced the risk of early death in many ways. The discovery of antibiotics, the understanding of the risks of smoking, the development of vaccines for many diseases have all contributed to reducing the health risks that we face through life. We now understand that the choices we make can affect our health, and that includes diet. The media continually brings medical research to our attention telling us how our dietary choices may be harming or helping us. It's sometimes hard to understand just how much we should allow this kind of research to change our lives.

For example, in 1989 it was discovered that the consumption of grapefruit juice can increase the uptake of most drugs.[98] It was subsequently found that the juice can also increase the concentration of the female hormone oestrogen in blood serum, by up to 30 per cent. Since higher levels of oestrogen are known

---

[98] This is why, for example, if you have been prescribed statins for reducing cholesterol, you are also advised not to drink grapefruit juice.

to be associated with higher levels of breast cancer, scientists at the University of Southern California and the University of Hawaii conducted a study[99] on post-menopausal women to investigate the effect of consuming grapefruit on the risk of developing breast cancer.

The study, which involved over 50,000 women, 1,657 with breast cancer, found that there was indeed such a risk. The headline result was that, for those who ate on average a quarter of a grapefruit a day, the risk of developing breast cancer was 30 per cent greater.[100] That sounds rather worrying. But 30 per cent greater than what? To understand how important that +30 per cent was, we first need to know a baseline figure.

In the study, the proportion of women who consumed *no* grapefruit but had breast cancer was around 3.4 per cent. Here's what that looks like in a simulated population of 10,000 women, of whom around 340 would, on average, be affected. The cells coloured red represent breast cancer cases.

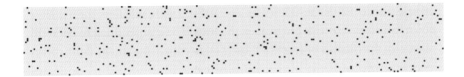

Now for grapefruit-eating women, we must increase the risk by 30 per cent to reach a proportion of around 4.4 per cent. In this simulation, there are 446 cases. The red cells are the baseline level, the blue cells are the additional cases associated with consuming grapefruit. Our example looks like this.

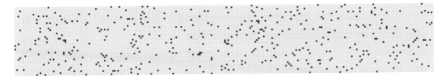

If we collect those red and blue cells together, you can get a better idea of the proportionate increase:

[99] 'Prospective Study of Grapefruit Intake and Risk of Breast Cancer in Postmenopausal Women: The Multiethnic Cohort Study', K. R. Monroe, S. P. Murphy, L. N. Kolonel, M. C. Pike, 2007.
[100] How sure were the scientists? 95% sure that the increased risk was between 6% and 58%.

The point of this is not to persuade you of the hazards or otherwise of eating grapefruit if you are a post-menopausal woman. The point is to show one way of understanding the risk. It may not be so helpful to know simply that the risk increases by 30 per cent. To get an idea of the importance of that increase, you first need to know what the baseline is, to understand if the overall impact is worth worrying over. If the percentages can be translated into case numbers, it provides another way to think about the probabilities. In that way you can make your own judgement as to what level of risk you'd be prepared to accept in return for your daily glass of grapefruit juice.

Screening

The value of health screening seems obvious. If there is a clinical test that can identify that certain people are at greater risk of some medical condition, be it a genetic condition, a cancer, heart problem, or anything else, then surely it makes sense to use that test? It seems logical that if you could filter the numbers down to a smaller at-risk group, then that would be a smart thing to do. You'd maximize the chances of successful early intervention. But things don't always work that way.

In 2013 in the United Kingdom, a screening process for dementia was proposed. The aims of the programme would be to identify patients at clinical risk of dementia and to offer follow-up action. It seemed a good idea: early intervention in dementia is generally thought to be a good thing. But the proposal and the subsequent implementation attracted controversy. For a start, people do not like to admit to failings, still less failings associated with deterioration of their mental faculties, and there was considerable resistance to the suggestion of senility implied by being selected for testing.

But even the mathematics was somewhat problematic, and exacerbated those human factors. The expected rate of dementia among the target group was 6 per cent. The test proposed was only moderately effective: it was expected to detect two-thirds of those who genuinely were at risk ('true positives'). But there would also be false positives: people who met the criteria of the test, but not because of dementia. The estimated false positive rate was about one-quarter. Working through the numbers, out of 100 people tested, 6 might be expected to genuinely be on the way to dementia and the test would identify, on average, 4 of them. Of the 94 without dementia, around a quarter of them, 23, would be falsely identified by the test. Overall, then, 27 might be flagged for follow-up action, and of these, the false positives would outweigh true positives by 6 to 1.

If the false positives came at no cost to the service or the individual, there would be little issue here: the screening tests could simply be a filter. Instead of dealing with 1 in 16 people who might have dementia, after screening we have 1 in 7. But the false positives are not without cost. Six out of seven of those followed up would have worried needlessly, and managing those follow-ups would have used resources which might have been used in other ways. There is a balance to be struck.

For calculations like this, the numbers that matter most are the effectiveness of the test and the relative prevalence of the condition in the population. The arithmetic itself means that even very accurate screening tests are of dubious value if the condition is rare. Suppose a condition occurs in just 1 in every 1,000 people, and suppose that for this condition we have a 99 per cent accurate screening test: only 1 time in 100 does it fail to detect the condition when it genuinely is there, and only 1 time in 100 does it falsely report the presence of the condition when it is absent. That seems a very effective test. Consider, though, what happens when the test is administered to 100,000 people:

Of the approximately 100 who have the condition, on average 99 are detected. Only one is missed as a false negative. Of the 99,900 who do not have the condition, on average 98,901 are correctly given the all clear, but 999 are wrongly picked up as positives. Even with such an effective test, the chance that a positive test correctly points to a positive diagnosis is less than 1 in 11. For every life potentially saved by the screening, 10 people will be subjected to the worry, stress, and possible health consequences of the follow-up. The judgement of where the balance of benefit and harm lies is of course a medical one, but it makes sense for it to be informed by an understanding of the probabilities involved.

Unnatural deaths

Most people die after leading long lives, when their bodies start failing from the effects of age. Others may die early from disease, and we feel these losses keenly. But some of the causes of death that we worry most about are those that we call unnatural.

The Office for National Statistics in the United Kingdom collects and publishes data collected from death certificates in England and Wales.[101] Of 3,137,759 deaths recorded for the six years 2013 to 2018 inclusive, 124,781 (fewer than 4 per cent) of them were due to unnatural causes.[102] Here are the top five categories within that list.

---

[101] It's a quirk of history, geography, and government that Scotland and Northern Ireland are *not* united with England and Wales for these purposes.

[102] Any analysis of cause of death will vary from place to place: territories with high crime rates and ongoing war or civil strife will not match these statistics.

Rank	Cause	Average/year	% of Unnatural deaths	% of all deaths
1	Falls	5,274	25.4%	1.01%
2	Intentional self-harm	4,090	19.7%	0.78%
3	Accidental poisoning[103]	3,082	14.8%	0.59%
4	Other injuries[104]	2,502	12.0%	0.48%
5	Land transport accidents	1,623	7.8%	0.31%

As in Graunt's London, cause of death is not always easy to determine. 'Falls' is at the top of this list. But every fall is caused by something, and determination of cause is not always clear-cut. We might expect that a proportion of those falls will be the consequences of failure of some part of an elderly body: a moment's dizziness or a wobbly knee.

Here's another selection of categories that are not in the top five, but which are of some interest in thinking about unnatural deaths and whether or not we should worry over them.

Rank	Cause	Average/year	% of Unnatural deaths	% of all deaths
10	Assault	290	1.39%	0.06%
12	Accidental drowning	220	1.06%	0.04%
14	Forces of nature[105]	64	0.31%	0.012%
16	Burns	23	0.11%	0.004%
18	Air (and space) transport accidents	18	0.09%	0.003%
19	Medical complications	18	0.08%	0.003%
23	Legal intervention and war	2	0.01%	0.0003%

---

[103] This includes drug overdoses and excessive alcohol consumption as well as deaths from carbon monoxide inhalation.

[104] That is to say, uncategorized injuries, and injuries other than those due to falls, mechanical forces, drowning, suffocation, electricity or radiation, smoke, burns, venom, forces of nature, poison, or overexertion, travel, or privation. This looks like a catch-all category that catches more cases than it should.

[105] This includes, among others, lightning, earthquake, and volcanoes as sub-categories.

What does 'cause' mean? The 'proximate cause' of death is the actual mechanism of death. If someone is murdered in a shooting, the proximate cause of death might be loss of blood leading to brain death through oxygen starvation, but that is just the end result of a chain of causation. The blood loss might be caused by injury to bodily organs; the injury due to a bullet. The firing of the bullet is the result of an action resulting from an intent to murder. Which of those is the cause of death? In our table it would appear under the single heading of Assault.

Here is a visualization of these chances of unnatural deaths. The 10,000 cells shown represent 10,000 deaths: the grey background cells are the natural deaths, while the various coloured spots represent the various kinds of unnatural deaths, 375 of them in this example simulation.

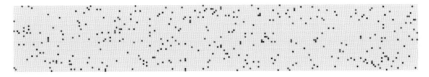

(red = falls, green = self-harm, blue = accidental poisoning, purple = other injuries, blue-green = road accidents, olive = other)

Here's another way of showing those same chances. It's the same grid, with the same number of coloured cells, but here they have all been gathered together:

(red = falls, green = self-harm, blue = accidental poisoning, purple = other injuries, blue-green = road accidents, olive = other)

Unnatural deaths are just not that common. In terms of micromorts, the risk of unnatural death in England and Wales is 0.8 of a baseline risk of 24 micromorts per day.

It is hard to think of anything more valuable than a day of life lived in good health and with the resources to make the most of that time. We rightly worry about lives cut short. But if worrying about things can lead to improvements, then the worrying may be worthwhile.

Risks of travel

It's a conversational cliché: 'Thousands of people die on our roads every year: who worries about doing anything about *that*?' Well, some people have been worrying about that. Despite the fact that we drive more than ever before, driving has become much safer over the years. Over a 16-year period, from 2004 to 2019, road traffic deaths in the United Kingdom declined from around 3,000 per year to less than 2,000 per year. The picture from the United States is similar: through four decades since the 1970s, road traffic fatalities reduced by two-thirds, and the number of injuries went down by more than half.

Road deaths are a small proportion of all deaths in the United Kingdom. On average, 30 out of 10,000 deaths. Here's a visualization of that proportion:

What about air travel? Many people fear flying, and it's not hard to empathize with that fear: modern air travel is a triumph of engineering, but the achievement of getting a jumbo jet into the air and keeping it there seems almost miraculous. Flying is not just an engineering miracle: modern air travel is also a triumph of safety management. Perhaps precisely because flying seems so unnatural, the airline industry has long known that addressing safety concerns is critical.

2017 was the safest year ever for commercial scheduled air travel. In that year, for the first time since commercial flights began, there were no fatalities on commercial scheduled jet plane flights. Please note, the wording there has been very carefully chosen. In 2017 there were indeed some deaths involving air travel, but they were on chartered (not scheduled) flights or involved planes other than jets (13 in turboprops). Nonetheless, 2017 remains the safest year on record for flying.[106] And the number for 2017, while something of an outlier, is still in line with an overall trend of increased safety over a period of decades.

There has been a remarkable reduction in the number and rate of fatalities arising from aviation: since 1970 (the year the Boeing 747 came into service) the decrease in the fatality rate represents a 54-fold lower risk over around 50 years.

---

[106] By contrast, in 2016 there were 271 fatalities; in 2018 there were 534.

What lies behind this level of safety improvement? As with car safety, part of the answer lies in technology. It may seem that the planes that carry us today are largely the same as when the jumbo jet was first flown over 50 years ago in January 1970, but this is not so. The technology that makes us safer is not always easy to see: new materials used in engines, better cockpit electronics, better sensors, navigation and air traffic control technology. Then there are human factors: better training and more realistic flight simulators contribute to the reduction of risk. But there may be an additional factor that lies behind all of these improvements.

In October 2018 a Boeing 737 Max, Lion Air Flight 610, crashed into the Java Sea, resulting in 189 deaths. Then in March 2019, just five months later, Ethiopian Airlines Flight 302, the same model of plane, came down six minutes after taking off from Addis Ababa: 157 people died. In both cases, it seemed that pilots were unable to control the plane. Speculation as to the cause has focused on the control software. Intense discussion and criticism of the design of the plane, the software, and the management of the company has followed, and the fleet of 737 Max planes was grounded, only recertified in the United States in November 2020. Paradoxically, the intensity of this reaction illustrates one of the reasons why air travel is now so safe, and it lies in something that Nassim Nicholas Taleb has called 'antifragility'.

This is about the appropriate response to failure. The reaction to the 737 Max crashes has been dramatic. The planes were grounded, and the model needed recertification before being allowed to fly again; the cost to Boeing was huge. This kind of response is not unusual. Through the decades, in part because plane crashes are so visible and because fear of flying is widespread, the airline industry and their regulators have responded to air accidents vigorously, often grounding fleets for extended periods. Exhaustive investigations follow, where every attempt is made to pinpoint the cause of the incident. This is followed up with remedial action. As Taleb explains, this means that crashes have the net effect, in the long run and on average, of making flying safer. Failure strengthens the system: not always through dramatic innovations, but often through small, incremental improvements. To use his term, the system is antifragile. In the case of the airline industry, it has made flying 50 times safer over the last 50 years.

The proportion of deaths due to flying shown in the table at the start of this section is so small that it's hard to form a clear idea of it: less than 1 death out of every 10,000 is related to flight. At this point I might have shown a visualization of that risk, but there's no point: 3 deaths out of 100,000 is too small to easily visualize.

Natural disasters

In England and Wales from 2003 to 2018 the proportion of deaths attributed by the Office of Natural Statistics to 'Forces of Nature' was just 0.012 per cent. Again, too small to visualize effectively. 'Forces of Nature' include storms, lightning strikes, earthquakes, and volcanoes. The frequency of these natural disasters is substantially a matter of geography: England and Wales certainly see their share of storms, but there have been no deaths from volcanic activity in historical times. However, things are not always so quiet on the other side of the planet.

Whakaari, also known as White Island, is the tip of a large, mostly submarine volcano, 30 kilometres off the coast of New Zealand, in the Bay of Plenty. On 9 December 2019 it erupted. 47 people were on the island at the time. It was announced later that day that 5 were known to have died, 34 people had been rescued from the island, and 8 people were still missing. In the weeks that followed, 14 more died from their injuries, and the 2 whose bodies had not been recovered were declared dead, bringing the total number of fatalities to 21.[107] How common are incidents like this?

In 2017 vulcanologists from the University of Bristol compiled a database of worldwide volcano-related fatalities from the historic record stretching back to the year 1500. From their data, in the 10 years from 2007 to 2016 inclusive, there were 554 deaths, worldwide, from volcano eruptions. As you might imagine, the numbers of deaths each year is highly variable. Here is a summary for six decades from 1950 to 2009:

Decade	1950s	1960s	1970s	1980s	1990s	2000s
Deaths	3,800	2,166	1,142	29,557[108]	3,461	1,469

As with volcanoes, so it is with earthquakes: some places are more exposed to risk than others. Earthquakes tend to happen along geological fault lines, those places where the plates that make up the Earth's crust touch each other, and are

---

[107] 'Volcanic fatalities database: Analysis of volcanic threat with distance and victim classification.' S. K. Brown, S. F. Jenkins, R. S. J. Sparks, et al. *J Appl. Volcanol.* 6, 15 (2017). https://doi.org/10.1186/s13617-017-0067-4

[108] This extremely large number contains an estimated 24,000 deaths resulting from an eruption of Nevado del Ruiz, Colombia, in 1985. The majority of deaths in this incident were due to lahars, mudflows having the consistency of wet concrete, that were created when pyroclastic flows melted snow and ice at the top of the mountain.

typically pushing together, or pulling apart. But taking the Earth as a whole, earthquakes happen with a frequency that is quite well described statistically. Small-scale earthquakes happen very frequently, and mostly pass unnoticed except by seismologists. And the relationship between relative frequency of earthquakes and their size is quite neat. Approximately, an earthquake that is one magnitude bigger than another will involve 10 times the shaking, and happen $1/_{10}$ as often. Roughly speaking, a magnitude eight (M8) earthquake happens about once a year somewhere in the world. That's a very large earthquake, and would make the newspaper headlines globally. An M7 earthquake, on the other hand, involves $1/_{10}$ of the shaking; these will tend to happen around 10 times a year. And so the pattern continues: an M6 earthquake will be expected around 100 times a year, perhaps 2 a week, somewhere in the world.

Asteroid strike

The generally accepted explanation for the massive extinction event that occurred at the end of the Cretaceous period is that an asteroid over 10 kilometres in diameter struck the Earth. If that were to happen again, it would likely trigger similar devastation. Should we be worried?

The good news is that such events have been calculated to happen only very rarely; on average, once in 100 million years. It's even better news that astronomers have been cataloguing and tracking the objects that stand a chance of striking us. Space is big and mostly empty, but there are still a lot of small rocks floating around out there. The asteroid-hunters believe there are around one million Near Earth Objects that could do serious damage on Earth (those that are 30 metres in size or larger). Such objects strike the Earth once or twice in every hundred years, on average. The bad news, though, is that to date only about 20,000 out of the million have been spotted, although that number disproportionately includes those capable of doing the most damage. It's thought that about 90 per cent of Near Earth Objects which are bigger than one kilometre are now being tracked. Of these, the one reckoned to have the biggest possibility of actually striking the Earth has been given the label 2000 SG344. It's 37 metres across (about $1/_{300}$ the size of the rock that killed the dinosaurs) and it's estimated that it has a 1 in 1,100 chance of impact in 2071. If it does hit an inhabited area, it will certainly cause deaths, but it is not large enough to trigger mass extinctions.

Should we worry about these objects? For myself, I'd lose no sleep over them, but I am glad that someone, somewhere, is keeping track.

Murder most foul

A 2018 report from the United Nations Office on Drugs and Crime reports that for the year 2016, there were around 450,000 murders around the world, a rate of 62 per million people. This figure, though, conceals enormous variation country by country. Here is a selection of country statistics, from high to low:

Country	Murder rate per million per year
El Salvador	618[109]
South Africa	359
Brazil	305
Colombia	249
Democratic Republic of the Congo	136
Nigeria	99
Russia	92
Afghanistan	62
Ukraine	62[110]
United States	53
Argentina	51
Pakistan	42
India	32
Canada	18
Israel	14
United Kingdom	12
Sweden	11
Germany	10
Italy	7
China	6
Indonesia	4
Japan	2

[109] About 10 times the world average.
[110] Matches world average.

In the United Kingdom, according to the table of unnatural deaths, Assaults ranked tenth, with an average of 290 deaths per year[111]—that is, 0.06 per cent of all deaths over that period. Here's what that rate looks like, as a proportion of 10,000 deaths.

Murders are widely reported in the news and other media, and the wide reach of such coverage means that we are disproportionately aware of such deaths. This is particularly the case in regard to child deaths: there is a widespread perception that we live in dangerous times. In fact, in recent times, the rate of child murders (age of victim under 16) in the United Kingdom has been quite consistent, at around 50 per annum, less than 1 per million of the population as a whole.

War and terror

This chart from the excellent OurWorldInData.com shows differing estimates of conflict deaths in recent years:

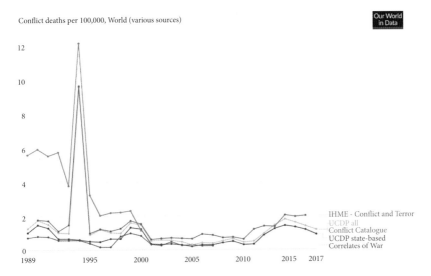

Barring the prominent peak in 1994, which represents the genocide that took place in Rwanda that year, the level of conflict deaths worldwide in recent years

[111] Not all deaths from assault are murders.

has hovered around 1 to 2 per 100,000. Compared to murders, that is about one-quarter of the murder rate. This worldwide view is only relevant, though, when looking at the biggest of big pictures. If one looks at different parts of the world, the rates vary: conflict deaths, unsurprisingly, mostly occur in areas of conflict.

Also from *Our World In Data* is this chart of deaths from terrorist activity:

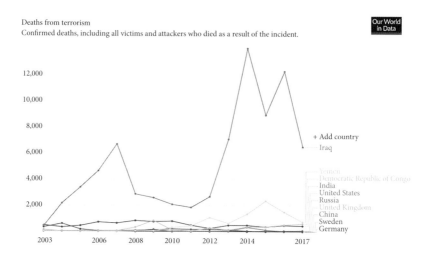

Deaths from terrorism
Confirmed deaths, including all victims and attackers who died as a result of the incident.

As you might expect, terrorist activity also varies greatly by country. If you were living in Iraq in recent years, with a population of 40 million, you'd have experienced a level of terrorism causing roughly 10,000 deaths per year, equivalent to 25 per 100,000 people living there. In Germany, on the other hand, there were only 37 deaths in total over the 15 years shown, a rate of 0.003 per 100,000— one ten-thousandth of the risk.

For a nation like the United Kingdom, or Germany, or the United States, despite the attention paid to every incident, the rates of death from murder, conflict, and terror are really too small for us to properly appreciate. They are far less common than they are made to seem, despite the media attention they attract. Of course each individual death is a tragedy. But we're up against the **individual–collective** duality again: what is unfathomably tragic at an individual level is insignificant at a collective level. Should we worry about murder and terror? Personally, I am happy that I live in a society that takes reasonable but not excessive measures to protect me from those risks, but I do consider that government spending might be better allocated to health services, which would protect far more people.

So, why worry?

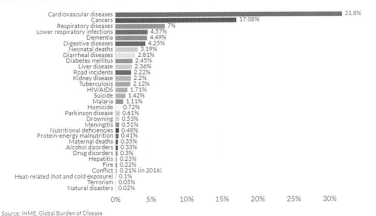

Share of deaths by cause, World, 2017

Data refers to the specific cause of death, which is distinguished from risk factors for death, such as air pollution, diet and other lifestyle factors. This is shown by cause of death as the percentage of total deaths.

Source: IHME, Global Burden of Disease

As the chart shows, deaths from unnatural causes form a very small proportion of all deaths, worldwide. If worry can be turned to useful action, improved safety from infections or accidents, that's useful worry. But when it comes to worrying about being the random victim of violence, it seems to me that there are more important things in life.

Chance need not always mean risk: it can also mean opportunity. And in the next section of the book, we will look at some of the ways in which chance can be a force for making the world, and our lives, better.

Answer to:
How long can your luck last now?

When there was just one joker in the pack, we reasoned that it could be in any position with equal probability. This variation, though, needs a little calculation. We're seeding the pack with two jokers, and so what matters is the location of the joker that appears earliest in the pack. We can see that it's impossible for the earliest joker to be the last card in the pack, and it is very unlikely that it is the second-last card. In fact, the further towards the end of the pack, the less likely is the first joker to appear there.

The way to calculate this is to take it a step at a time. Before flipping the first card, there are 52 safe cards out of 54, so the chances of surviving the first card flip are $^{52}/_{54}$. Having removed that first safe card, the chances of surviving the second card flip are 51 out of 53, or $^{51}/_{53}$. To calculate the chance of surviving both flips, we must multiply together: $\dfrac{52 \times 51}{54 \times 53}$. The pattern continues. The chances of surviving three card flips are: $\dfrac{52 \times 51 \times 50}{54 \times 53 \times 52}$. Now the sharp-eyed reader will notice that in that calculation, there is a factor of 52 both above and below the division line, which cancel each other out and make the calculation easier. This pattern continues, so if we continue the logic, we find that the chance of surviving 15 flips is $\dfrac{39 \times 38}{54 \times 53}$, which comes to 51.8 per cent, while the chance of surviving 16 flips is $\dfrac{38 \times 37}{54 \times 53}$ which is 49.1 per cent.

So, you have better than even odds of dealing 15 cards without seeing the first joker, but worse than even odds if the target is 16. Insist that the stranger gives you a target of 15 or fewer.

There are plenty of jokers in our lives: if one cause of death doesn't kill you, then another surely will. Fewer early deaths means that the causes of death that predominate later in life, such as cancer, take a greater and greater share of mortality, and calculating the effects of combining multiple risks is seldom trivial.

PART 3

Happy Accidents

Iannis Xenakis: Maker of mathematical music

Since antiquity the concepts of chance (tyche), disorder (ataxia), and disorganization were considered as the opposite and negation of reason (logos), order (taxis), and organization (systasis). It is only recently that knowledge has been able to penetrate chance and has discovered how to separate its degrees.

Iannis Xenakis, *Formalised Music*

Iannis Xenakis, the avant-garde composer who pioneered techniques ranging from sampling to computer-driven composition, was born in Romania in 1922, of Greek heritage. Both his parents were musical, but it was from his mother, who died when he was five years old, that he believed his love of music came. Music was not his first choice of career. He passed the entrance exams for the National Technical University of Athens, where he planned to study engineering and architecture, but his plans were disrupted by war when in October 1940 the Italians invaded Greece. Xenakis became involved in armed resistance, and in the civil war that followed the departure of the occupying Axis forces in 1944, Xenakis, now enlisted with a communist student group, was injured in street fighting against British tanks and lost an eye.

In 1947 Xenakis graduated in civil engineering from the Polytechnic School in Athens, and immediately fled Greece for fear of political repercussions, travelling via Italy to France. He found himself a junior position in the studio of architect Le Corbusier. Eventually, he started collaborating with Corbu himself and was given responsibility for solo projects. The most celebrated of these was his design for the Philips Pavilion for the international exposition in Brussels in 1958, which made use of unusual mathematical shapes, specifically hyperbolic paraboloids, all swooping smooth curves.

At the same time, Xenakis was studying music theory, and composition in particular.

By the 1950s his teacher was the innovative composer, Olivier Messiaen, famed among other things for incorporating transcriptions of birdsong into musical pieces. Xenakis asked Messiaen whether he should resume the study of conventional musical elements of harmony and counterpoint. 'No,' replied Messiaen, 'you are almost 30, you have the good fortune of being Greek, of being an architect and having studied special mathematics. Take advantage of these things. Do them in your music.'

And he did. Much is made of the links between mathematics and music, but often the connections identified are rather superficial. In essence, both are abstract arts and they both use specialized notation that appears daunting to the uninitiated. Music depends on physics, and physics is deeply mathematical—notes that are musically consonant tend to have frequencies that bear neat mathematical ratios to one another. But the way in which Xenakis introduced mathematics into music was anything but superficial; perhaps because of his architectural background, he was accustomed to using mathematics to inform his creativity. For Xenakis, mathematics would not be merely a starting point for composition, but a genuine attempt to explore mathematical ideas and how an audience can experience those ideas through his compositions; among these were ideas drawn from probability theory. In some of his works, the performers are instructed to make sounds randomly, with no explicit score, but adhering to statistical guidelines as to density (how many sounds per second to make, on average), pitch, and duration. In this way, a 'cloud' of sounds is created, with specified statistical properties but no definitive set of individual notes. The collective effect is what matters.

Xenakis was a writer, too, his most notable book being *Formalized Music: Thought and Mathematics in Composition*. He died in Paris in 2001 at the age of 78, by then recognized as a leading composer of art music.

Happy Accidents

The new always happens against the overwhelming odds of statistical laws and their probability, which for all practical, everyday purposes amounts to certainty; the new therefore always appears in the guise of a miracle.

Hannah Arendt

Serendipity

In 1557 a book by Michele Tramezzino was published in Venice, recounting an ancient Persian fairy tale concerning three princes accused of stealing a camel. The princes, claiming innocence, are charged to prove it by finding the missing camel. While they exercise Sherlockian powers of reasoning in their hunt for the camel, their efforts are fruitless, at least as far as the camel is concerned. But along the way they repeatedly make accidental discoveries unrelated to their main quest. In the end the camel is indeed found, though not through their efforts, but simply by chance. Still, their innocence is proved and their lives are spared.

The setting of the story is the island of Sri Lanka and the tale is called 'The Three Princes of Serendip' (the old Persian name for Sri Lanka was Serendeppo). Once translated, the story became quite well known throughout Europe. In 1754 Horace Walpole (the author of the first Gothic novel, *The Castle of Otranto*, and the son of prime minister Robert Walpole) made reference to it in a letter to one Horace Mann. In this letter he coined the word *serendipity* to mean the quality of making happy, but unsought-for discoveries by chance, and it entered the English language. In a recent poll of favourite English words, *serendipity* topped the list. A beautiful word for a charming and happy concept.

hap (n.)

> (c.1200) 'chance, a person's luck, fortune, fate'; also 'unforeseen occurrence'. From
>
> *happ* (Old Norse) 'chance, good luck'. From
>
> *hap-(Proto-Germanic root). From
>
> *kob- (Proto-Indo-European root) 'to suit, fit, succeed'

The meaning of 'good fortune' in English is from the early thirteenth century. And from *hap* we get *happy*, of course, but by contrast not only *unhappy* but also *hapless*. And *happen* and *haphazard* and *mayhaps* and *perhaps*. And those last three meanings bring us right back around to chance.

'No news is good news', we're sometimes told. That also makes sense when turned around: good news is no news. Why is the evening news always filled with disasters? Why can't they report some good news for a change? Because good news is no news. Loosely interpreted as a statement of information theory, that's true. Good news, by and large, means hearing that projects have gone to plan, that things are normal, stable, that everything is on course. In information theory terms, good news is often a communication of the most likely outcome, and that is the message that bears the least information content. So, most of the time, good news is no news, and most news is bad news.

That is how we usually experience the effects of chance. Chance events tend to be disruptive. When randomness strikes, it comes as a surprise, and most often the surprise is unwelcome. But not always. Sometimes chance works to our advantage: every now and again a happy accident occurs. We can see chance as an essential driver of creativity: in the sciences and in the arts, chance occurrences and random alignments lie behind many discoveries, innovations, and creations. Even mistakes have their role. We stumble, and find that we have tripped over half-buried treasure. Sometimes literally.

CHAPTER 10

Chance Discoveries

A discovery is said to be an accident meeting a prepared mind.

Albert Szent-Gyorgyi

Rare and valuable

With collectibles, rarity can be as important as inherent value or age. Do you know, or can you guess, which of these coins is the most valuable? (It was sold at auction for more than twice the price achieved by any of the others.)

- A US 'liberty head' nickel, minted in 1913 and sold in 2018
- A US 'flowing hair' silver dollar, minted in 1794, sold in 2013
- A US 'Birch' cent, minted in 1792 and sold in 2015
- An Ummayad Caliphate Gold Dinar, minted in 723 and sold in 2019
- An ancient Greek silver decadrachm, minted around 400 BCE and sold in 2012

Answers at the end of the chapter.

Treasure trove

On 21 December 2014, at a rally of metal detectorists in a field near the village of Lenborough in Buckinghamshire in England, one of the treasure-hunters found a lead container buried half a metre deep. Peering through a hole in the lead, Paul Coleman spotted a glint: a shiny coin. That coin turned out to be the

What are the Chances of That? How to think about uncertainty. Andrew C. A. Elliott,
Oxford University Press. © Andrew C. A. Elliott 2021. DOI: 10.1093/oso/9780198869023.003.0010

first of 5,251½ coins. The fraction? One of the coins had been cut in half before burial. The club's founder, commenting on the pristine nature of the silver coins, said he suspected that they had, prior to their discovery this century, only ever been handled by two people: the one who made them and the one who buried them. The coins appeared to date from the early eleventh century, and were minted in the reigns of Æthelred the Unready and Cnut. The estimated value of the Lenborough hoard is in excess of one million pounds.

It's a staple of children's stories: buried treasure. And it's more than a story. In the United Kingdom in 2017, more than a thousand finds of buried treasure were reported. Most of the finds were of objects, but 18 per cent were of coins. A quarter were from the medieval period, and a small number dated back to the Bronze Age. The finds of coins were mostly from the Romano-British period. But don't think that if you stumble over a hoard of treasure you automatically get to keep it. In the United Kingdom, treasure[112] belongs to the Crown (the State), and must be reported. You, as finder, and whoever is the owner of the land where it was found may at the discretion of a regulatory body be given a reward.

What counts as treasure? The Treasure Law that applies in England says that treasure is a find at least 300 years old of:

- Coins of any metal, where there are 10 or more found together
- Coins of at least 10 per cent precious metal, where two or more are found together
- Objects with metal bits that are at least 10 per cent precious metal

So jewels, ivory, or amber are not treasure in and of themselves, but only when fitted to precious metal mounts; nor does a single gold coin fall into the definition. Treasure according to English law must be reported when found, and must be offered to a museum (after which a reward may be granted). If no museum is interested, the finder can keep the hoard. In 2017 in the United Kingdom (excluding Scotland) almost 400 finds went to museums, with around 700 returned to the finder or landowner.

In Scotland, the law is different (and older). The term used for a find is treasure trove (from the French *trouve*, meaning 'found'), and it need not be metal: any object considered to be significant, regardless of its age or composition, may be claimed as treasure by the Crown. In 2017 over 150 such finds were

---

[112] In this usage, 'treasure' is a legally defined word.

allocated to museums, and members of the public were awarded around £2 million in compensation.

In the United States, the law of treasure trove varies from state to state, but the general principle is that it must be gold or silver, or paper money (deemed equivalent to gold or silver), that has been hidden long enough (several decades) for the original owner to be unlikely to claim the find. In most states, the finder can keep the treasure.

Unearthing the past

In 1709 in the town of Resina, on the bay of Naples, workers assigned to dig a deep well for a monastery came across high-quality marble statuary. These statues turned out to be part of the theatre of the affluent Roman town of Herculaneum, destroyed when Mount Vesuvius erupted in 79 CE. The town had been covered over with ash and rock to a depth of 16 metres, effectively sealing it off from the outside world for 1,630 years. Deep under the volcanic material, workers found small charred objects, some of which they used as charcoal briquettes. They turned out to be papyrus scrolls from the library of a wealthy Roman's villa. The destructive heat of the volcanic material had, somewhat paradoxically, preserved them in a fashion, and protected them from the usual fate of papyrus: water damage and bookworms. When discovered, very little could be done with those papyri, but since the 1990s it has been possible to read some of the fragile volumes using multi-spectral imaging.

More than a century earlier, in 1592, workers had come across ancient walls decorated with paintings while digging a canal in an area called Civita, a little further along the bay of Naples. The ruins there were noted and re-covered. It was only in 1748 that further deliberate work was undertaken to uncover the ruins, and it was not until 1763 that the ruins at Civita were definitively identified as the Roman town of Pompeii. Because Herculaneum and Pompeii were buried so suddenly, and so completely, they have become invaluable treasures: snapshots of Roman life captured on a single day in 79 CE. An unpredictable chance event both destroyed and preserved them, and in the end they were both found by accident.

Much of what we know of the past comes from chance discoveries. The caves at Lascaux, filled with astonishing prehistoric paintings, were discovered when Marcel Ravidat was searching for his dog, Robot, who had been chasing a rabbit. The Terracotta Army at Xi'an in China was discovered in the course of digging a well. The Rosetta Stone, the key to understanding ancient Egyptian hieroglyphic writing, was uncovered by members of Napoleon's army in 1799 working to reinforce Fort Julien, a few miles from the town of Rosetta, near modern-day Rashid in Egypt. A young boy searching for a stray goat on the shore of the Dead Sea, some thirty kilometres from Jerusalem, entered a cave, and found what we now know as the Dead Sea scrolls.

Our knowledge of the distant past is fragmentary and random. Which of the scrolls from the Villa of the Papyri in Herculaneum were saved and which were lost was a matter of chance. The past has presented us with only a handful of pieces from a thousand-piece jigsaw, and we are trying to make sense of the whole. The picture is undoubtedly incomplete: the question of how accurate it is, is probably unanswerable. And it is a matter for pure speculation to wonder about how different our picture would be, had a different set of random discoveries been made.

Scientific discoveries

One sometimes finds what one is not looking for.

Sir Alexander Fleming

Robert Koch (1843–1910), one of the founders of modern bacteriology, was a pioneer in the development of the germ theory of disease. As part of his work at the Imperial Health Office in Berlin, he needed to grow pure cultures of various bacteria. At the time, this was done in nutrient broths that were difficult to manage. One day he happened to notice a culture of a single strain of bacteria growing on the surface of a boiled potato. This chance observation led him to realize that a solid nutrient medium would work much better, and he started using potato slices routinely as a medium for growing cultures. The flat, solid surface provided stability and was a more practical way of investigating the bacterial cultures. His laboratory colleagues refined this approach. Fanny Hesse suggested using agar gels instead of potato slices; Julius Petri suggested that for ease of observation with minimum contamination, the agar could be contained in shallow glass dishes with flat glass covers. These dishes are now ubiquitous and are known everywhere as petri dishes.

One thing leads to another: petri dishes themselves were soon to take centre stage in a chance discovery. On returning from holiday to his laboratory in St Mary's Hospital, London, in 1928, Dr Alexander Fleming discovered that some of his petri dishes had been contaminated by mould. Where the mould had grown, staphylococcus bacteria had died. The mould was *Penicillium*[113] *notatum*, and it took over a decade of work by many people to extract and refine the active ingredient, later named penicillin. Fleming wrote: 'When I woke up just after dawn on September 28, 1928, I certainly didn't plan to revolutionise all medicine by discovering the world's first antibiotic, or bacteria killer. But I guess that was exactly what I did.' Chaos, in the form of a messy laboratory, stands at the start of this chain, but it took a trained mind to see and understand what chance had created, and disciplined and methodical work to nurture the discovery to create a viable drug.

A chemical with formula $Pt(NH_3)_2Cl_2$, a compound of platinum, was known as Peyrone's Chloride, originally created in 1844 by Italian chemist Michele Peyrone. But in 1965 a biophysical chemist, Barnett Rosenberg, was conducting experiments on bacterial growth in electric fields, and he found that when he used platinum electrodes, the bacteria grew much larger. He traced this to a compound created when the platinum corroded: the same Peyrone's Chloride. On further investigation, he discovered that the chemical, now called cisplatin, could cure tumours in mice. It is now one of the most widely used cancer treatments.

Saccharin, the first artificial sweetener, was discovered in 1879, when researcher Constantin Fahlberg failed to wash his hands after a chemical spill, and the chemical transferred to his bread roll at dinner that evening: it tasted oddly sweet. The first commercial aniline dye, mauveine or mauve, was noticed by 18 year old William Henry Perkin when in 1856 he was trying to develop a synthetic version of quinine for the treatment of malaria. This list could be continued at great length: it could almost be said that chance is an essential component in the scientific process.

Many accounts of the historical development of the sciences centre on the people involved, the geniuses whose intellectual acuity equipped them to take the giant steps needed to advance our understanding of the world. The story that a randomly falling apple prompted Isaac Newton's breakthrough in developing a theory of gravity is almost certainly not true. Still, scientific progress is often portrayed as a series of landmark discoveries by exceptional figures: the likes of Archimedes, Newton, Maxwell, Darwin, Curie, Einstein. But these individuals were statistical outliers. Perhaps unusual intellectual prowess was necessary for breakthrough advances, but it was also necessary for the

---

[113] *Penicillium* is Latin for a painter's brush, so named for the appearance of the fungus.

circumstances of discovery to be right. The right person needed to be in the right place, with the right social credentials, at the right time, and to a great degree this alignment is a matter of chance. Would our world today be different if these great thinkers had never existed, or never been able to do the work that marks them out? Perhaps not so different. We know that Newton's work on calculus was paralleled by that of Leibniz, that some of Charles Darwin's ideas were shared by Alfred Russel Wallace. Maybe the ideas are more important than the discoverers.

The role of serendipity in science

The most exciting phrase to hear in science, the one that heralds new discoveries, is not 'Eureka!' but 'That's funny...'

Isaac Asimov

These anecdotes of how chance has played a role in scientific discovery are very pleasing. It's somehow comforting to know that for all the brilliance and hard work of scientists, sometimes success comes from a stroke of luck. But the part that chance discoveries play in science is not just as a source of entertaining stories. The question 'what were the chances of that?' for such serendipitous discoveries is unanswerable. But they are sufficiently common that we would be unwise to dismiss chance as a significant factor in science. More than that, it makes sense to consciously leave space open for random occurrences to happen and to be noticed.

This factor is significant enough in scientific research to merit study in its own right. In his paper 'Serendipity: Towards a taxonomy and a theory', Ohid Yaqub reports on his studies of hundreds of cases of scientific serendipity. He identifies four types of serendipitous discovery. His classification is based on two factors: the motivation and the outcome.

Walpolian serendipity: Targeted search solves unexpected problem. This is named by Yaqub for Horace Walpole, who first coined the word serendipity. Appropriately, this type of serendipity corresponds to Walpole's original meaning: discovery of things that the discoverers were not in search of. Here, the motivation is targeted (the researchers are searching for something specific) but the outcome relates to a different problem.

Mertonian serendipity: Targeted search solves problem-in-hand via unexpected route. This is named for Robert K. Merton, a sociologist of science, whose archive provided much material for Yaqub. Merton extended the meaning of serendipity to include those cases where the seeker finds the target of their search, but where the sought-after outcome is arrived at in an unexpected way. Charles Goodyear, searching for a way making a thermostable rubber, by accident combined rubber and sulphur on a hotplate, and thereby invented the process of vulcanization.

Bushian serendipity: Untargeted search solves immediate problem. This gets its name from Vannevar Bush, who led the US Office of Scientific Research and Development during the Second World War. This is where research with no specific aim in mind stumbles across a solution to a current problem. Saccharine's properties as an artificial sweetener were discovered by Constantin Fahlberg forgetting to wash his hands before eating and accidentally tasting chemical residue.

Stephanian serendipity: Untargeted search solves later problem. This is named for economist Paula Stephan, and relates to discoveries whose value is not immediately recognized, but which are picked up later when a problem arises for which the earlier discovery provides a solution. In 1903 French chemist Édouard Bénédictus dropped a chemical flask on the inside of which a film of collodion had formed. The flask broke, but the pieces remained connected. Six years later, when Bénédictus heard of a car accident involving injury caused by flying glass, he patented the idea of laminated safety glass.

Yaqub also identifies four mechanisms, some of which are typical of serendipitous discoveries:

- **Theory-led**: The serendipitous discovery may be inconsistent with conventional theoretical understanding, and so serves to develop the theory.
- **Observer-led**: The observer may be uniquely placed to make the discovery through their particular mix of tools, techniques, or attributes.
- **Error-borne**: Chance discoveries are often associated with errors such as breaches of protocol or accidental spillages. Yaqub quotes Max Delbrück referring to this as 'limited sloppiness'.
- **Network-emergent**: Sometimes the discovery needs an outsider's viewpoint. Yaqub cautions, though, that networks can also stifle serendipity by encouraging groupthink.

Should these considerations affect science policy? Are there ways to arrange matters to provide a fertile environment for fruitful accidents? Yaqub suggests so. Perhaps the role of chance in scientific discovery is too important to be left to chance.

Contingency or necessity?

contingency (n.)

> 'quality of being contingent, openness to chance or free will, the possibility that that which happens might not have happened'. From

> **contingent**, (late 14c), 'depending upon circumstances, not predictable with certainty, provisionally liable to exist'. From

> > **contingentem** (Latin) happening; touching'. From

> > > **contingere** 'to happen to one, befall, come to pass', originally 'to touch'.

Which of these archaeological and scientific discoveries were inevitable, and which might still be unknown, but for the intervention of chance? It's conceivable that Herculaneum could still be buried: it's a relatively small site, and the layer of rock and ash that covered the town was very thick. You can easily imagine a world where the Rosetta Stone was smashed to pieces before its value was recognized. It's hard to imagine that we would still not yet have decoded hieroglyphics. And, if not Fleming, surely someone else would have found the first antibiotics.

Perhaps all of these discoveries could and would have been made over the years by systematic, directed search. But, without the operation of chance, their discovery could have been delayed by years. No doubt the penicillium mould had been contaminating other petri dishes in other laboratories for years. But it took Fleming to notice what was happening and a team of scientists to turn that noticing into actual curing. So turning chaos into active, positive change is not simply a question of allowing sense to emerge from nonsense: it requires a filtering process, a mind alert to possibilities, a process of nurturing and refinement.

In Voltaire's satire *Candide*, Professor Pangloss holds that 'all is for the best in this, the best of all possible worlds'. Pangloss, with his easy optimism, is a figure of ridicule, but the view reflects a more serious attitude. Voltaire was satirizing the view of Gottfried Leibniz, and his attempt to account for suffering and evil in the world. Leibniz's argument comes down to this: that it would be irrational for an omnipotent God to create a world that was not exactly as intended. This creation must therefore be the best of all possible creations, and should be judged by its conformity to God's intentions, and not by how well it suits us. According to this viewpoint, the world is as it is not by chance, but because it must be this way.

But if we reject the idea that everything that happens is explicitly directed by an ever-attentive deity, then we must conclude that the world is some mixture of the random and the inevitable. A boulder is dislodged in a violent storm. It tumbles down a mountain and lands in a stream, whose course is then diverted. The land has been shaped by a random event: had the boulder fallen in a microscopically different direction, the local geography would have been different in a macroscopically noticeable way. And so the world is changed by chance.

Were the serendipitous discoveries described in this chapter inevitable, or merely what happened to happen: mere contingencies? It's hard to think that Newton's understanding of gravity would have remained hidden for very long. Even Einstein's radical reinterpretation of that understanding would surely have been found by someone else. It feels that the discovery of truths such as these was inevitable, even if the details are contingent, simply the way in which things happened to happen. The near-simultaneous invention of the calculus by both Newton and Leibniz shows that two contingent narratives can reach the same endpoint. But is the same true of the discovery of useful chemicals? Was penicillin inevitable? Probably. Was cisplatin? Are there other drugs out there, undiscovered? Unquestionably. Are there new discoveries being made? All the time.

We know that to work out the probability of a compound event, we must multiply the probabilities of every separate part. So the chance of our world being as it is now must be vanishingly small—the product of uncountably many probabilities, each of which is less than one, and which when multiplied together give a result that is close to zero. And yet this is the world we are in, with complete certainty. And there we have perhaps the most extreme example of the **hindsight–foresight** duality: that the chance of this present reality, the probability that we would be living in this precise world, is at the same time both a hair's breadth away from complete impossibility and yet is absolutely certain.

The discoveries that scientists make may or may not be inevitable, but the truths they discover are far from random: they are models of reality. In contrast, the works of artists may reflect equally deep truths, but they are undoubtedly contingent: simply a matter of how things happened to happen. Only Michelangelo could have sculpted *that* David, only Bach could have composed the Brandenburg Concertos. Our artistic heritage is a consequence of chance alignments: the creations of particular people in particular places at particular times. But artists are not mere pawns in a game of chance: as we will see in the next chapter, they can take control and deliberately seek to involve probability in their creative processes.

Answer to:
Rare and valuable

Here are the coins, ranked by how much they achieved at auction.

- A US 'flowing hair' silver dollar, minted in 1794, sold in 2013
  - The coin sold at auction for $10,016,875. This is the highest price (to date) ever fetched by a coin. The coin in question is the finest known example of the first dollar coin issued by the United States, which matched the Spanish dollar for size and weight. The 'flowing hair' is that of a portrayal of Liberty on the 'heads' side.

- An Ummayad Caliphate Gold Dinar, minted in 723 and sold in 2019
  - The coin sold at auction for $4,780,000. It weighs 4.25 g and is gold. Ironically, the name 'dinar' derives from the Latin *denarius*, which was a silver coin.

- A US 'liberty head' nickel, minted in 1913 and sold in 2018
  - The coin fetched $4,560,000 at auction. It was minted in very small quantities: only five examples are known to survive.

- An ancient Greek silver decadrachm, minted around 400 BCE and sold in 2012
  - This coin was sold at auction for $2,918,000. Decadrachm means 10 drachms, and this coin was minted in Agrigentum in Sicily.

- A US 'Birch' cent, minted in 1792 and sold in 2015
  - This coin was sold for $2,585,000. Named for its engraver, Robert Birch, it was one of the first coins minted by the United States, and is extremely rare.

Mixing it Up

To dare every day to be irreverent and bold. To dare to preserve the randomness of mind which in children produces strange and wonderful new thoughts and forms. To continually scramble the familiar and bring the old into new juxtaposition.

Gordon Webber

Fortune favours the bold?

'Let me get this straight. You've taken just the 13 diamonds from that pack of cards. I'm going to flip them over one by one, and for each one I reveal that is *not* the jack of diamonds, I get a dollar?'

'Yes, more or less. You put three dollars into the pot, and I will put nine in. Then for each card you turn over that is not the jack, we set aside one dollar for you. So long as Jack hasn't shown himself, you can end the game whenever you like, and claim your winnings. But if the jack of diamonds turns up, you lose it all. Well, you lose three dollars, which is what you added to the pot. If you're brave enough to keep going, and the jack happens to be the last card, you walk away with the whole 12 dollars.'

'But that would be pushing my luck. It would make more sense to stop before it got that far.'

'That would seem sensible. But it's your choice: you can play this as recklessly or as cautiously as you like. And remember that I've got a lot more to lose than you do.'

What are the Chances of That? How to think about uncertainty. Andrew C. A. Elliott, Oxford University Press. © Andrew C. A. Elliott 2021. DOI: 10.1093/oso/9780198869023.003.0011

He has a point, but it seems that you need a strategy. You have nothing to lose by seeing the first card, but once your winnings start piling up, each turn of the card is more risky, and there is more to lose.

So, what's your best strategy? And what's the chance you come out of this with a profit? Answers at the end of the chapter.

Did Amadeus play dice?

In 1793 the musical publisher J. J. Hummel put out a musical curiosity. It was a little parlour game, a *Musikalisches Wuerfelspiel* (musical dice game), that the publisher claimed had been devised by Wolfgang Amadeus Mozart. Having died two years previously in 1791, Mozart wasn't around to contradict the publisher, and his name has remained attached to the game ever since. Although there is no proof that the game was Mozart's invention, it's not a ridiculous idea. Found among his effects after his death were the makings of a musical game that worked along broadly similar lines.

The game is a way for amateurs to compose their own waltzes, using dice rolls to incorporate random elements and ensure that each waltz produced would be entirely novel—never heard before. Mozart provided 272 musical fragments, each of them a single bar of music, and he devised a set of rules for these fragments to be assembled randomly based on dice rolls. His rules narrowed the choice of fragment for any one bar, and so ensured that the strung-together random choices respected rules of harmony and form, and would result in a playable and harmonious (but perhaps not particularly inspired) waltz.

The waltz form he used is in three parts. The first is a *minuet* of 16 bars, which is followed by a *trio* of 16 bars, after which the original minuet is repeated. Each bar of the minuet is selected by rolling two dice and selecting one of 11 fragments (the spots on two dice can give totals from 2 to 12). For the minuet, then, with 11 possible choices for each of the 16 bars there are 11^{16} possibilities—a very large number! You may notice that for this minuet section the chances for each fragment are not equal—when two dice are rolled, a 7 is much more likely than a 2, for example. So those 11^{16} possibilities do not all have equal chances of appearing.

The generation of the 16 bars for the trio uses only one dice, and the selections are made from six fragments. That makes for 6^{16} possibilities, and these *are* all equally likely. Taken together, that means Mozart's little game could generate

$11^{16} \times 6^{16}$, or approximately 1.3×10^{29} distinct tunes.[114] If each waltz lasted two minutes, then to play through them all would take 37 trillion times the age of the universe.

Mozart (or whoever really did put the game together) constructed his atoms of music, his one-bar fragments, very carefully, so that whichever choices were made, when strung together, they formed a coherent harmonic structure. This meant that while at the level of a single bar the musical fragments were randomly selected, those random choices sat within a fixed framework of form and harmonic development, which ensured that the waltz made musical sense. This is a pattern we see over and over where randomization is used in music. Some elements remain fixed, and provide the listener with a degree of familiarity, while other elements are free to vary.

A waltz constructed using Mozart's dicey game would, with a very high degree of certainty, be unique. But the finished product would not betray its random origins. Once the dice-rolling had been done and the fragments had been assembled, Mozart's rules meant that the final product would always be a coherent, finished piece of music, with nothing particularly odd about it. Chance might be involved in its composition, but not in its performance.

Improvisation

Music has always embraced elements of chance, of improvisation, in performance. Even if one of the aims of a classical music performance is fidelity to the composer's intentions, individual performers inject something of their own character in the way they interpret those intentions, and so add an additional level of artistry. Sometimes improvisation is expressly allowed for: in a concerto it is conventional, towards the end of the movement, for the composer to mark a place for the solo instrumentalist to take flight in a passage of improvisation known as a cadenza.[115]

But no type of music has made improvisation its own in quite the way that jazz has. One of the joys of jazz, both for the listener and the player, is the space

---

[114] The second minuet is a repeat of the first, and so adds no further variation.

[115] It probably is a coincidence that this word, *cadenza*, shares its roots (Latin *cadens*, falling) with the word *chance*. *Cadenza* is related to *cadence*, which in musical terms means the final notes of a phrase of music as it falls away.

it allows for chance events and spontaneity.[116] In most jazz performances, there are explicit opportunities for improvisation. Whoever is soloing can play the music that arises in their mind, in the moment. The improviser's skill is to create something new that nonetheless fits within the established framework of time and harmonic development. It's a balance between freedom and structure. The improvisation, if honestly done, is unlikely to be a repetition of anything ever played before. Then too, in small jazz groups at least (big bands are a different matter), there is a fluidity and a flexibility in the way in which a piece is arranged (how long the piece continues, who takes solos, how the other members choose to support the soloists). These decisions are very often made entirely on the fly, and they make any performance a one-off.

Unlike in a classical performance, there is little value attached in a jazz performance to the reproduction of a perfect canonical rendition: generally speaking, jazz players value their own signature sound, and the sound of their group, more than the intentions of the person who wrote the music. The way jazz allows free rein within constraints reminds us a little of the Mozart game, in that the structure of the piece is taken as given, and it forms a framework for the players. For any given piece, the harmonic progressions will be known, and players will either respect the written harmonies or knowingly violate them for intended effect. Just as Mozart constrained the choices for each bar of music in his game so that the whole would make harmonic sense, a jazz soloist will improvise with the harmonic structure of the piece in mind. Even if she chooses to play outside that structure from time to time, she does so in the knowledge that she is overturning expectations. The jazz saxophonist Jerry Coker, in his book *How to Improvise*, discusses what it takes to build a compelling improvised solo:

> The listener is constantly making predictions; actual instantaneous predictions as to whether the next event will be a repetition of something, or something different. The player is constantly either confirming or denying these predictions in the listener's mind. As nearly as we can tell…the listener must come out right about 50% of the time—if he is too successful at predicting, he will be bored; if he is too unsuccessful, he will give up and call the music 'disorganised'.

So a jazz audience needs to be seduced with a blend of the expected and the unexpected. The jazz fan will have some idea of where your melodic line is

---

[116] This makes recordings of jazz performances somewhat paradoxical. The recording captures the spontaneity and at the same time entirely obliterates it. Can your astonishment at a flash of genius be fully recaptured when relistening to a recording? Is this an example of the **foresight–hindsight** duality?

heading and how your harmonies will develop. Striking the right balance between expectation and surprise is fundamental to a successful solo.

In jazz, most improvised solos span a few choruses of (typically) 32 bars each, and last a minute or two. Some jazz players, though, notably the pianist Keith Jarrett, take improvisation a lot further, and may spend an hour or more in concert, publicly performing entirely fresh material, developing a musical idea freed from any fixed plan, and guided only by their musicality. It's a remarkable experience to attend such a concert, where you know that what you are hearing has not been heard before, has not been played before, and will not be played again. It demands a degree of focused attention from the listener, and is a far cry from, say, a rock concert where the audience has arrived in the expectation of hearing the band's greatest hits.

A jazz performance puts chance on show a little more than a typical classical concert. There is a stronger sense that what happens in each piece is unscripted, that music is being made in real time, for that place and that time. Mistakes are embraced. There are plenty of opportunities for happy accidents and flashes of inspiration. But we are still some distance removed from a genuinely chance-driven performance.

Indeterminate music

Composer John Cage's most notorious work is his *2'33"*: 2 minutes and 33 seconds of silence. At the first performance in 1952, the pianist David Tudor marked the start of the piece by *closing* the piano lid over the keys. He then opened and closed the lid to mark the breaks between movements.[117] The intention of the piece is to give the audience the time, and the silence, to stop and listen, to hear the 'music' of the ambient noises around them, noises both in the auditorium and from the world beyond, the muffled hoot and grumble of street traffic, someone yelling, doors slamming. Cage talked of 'indeterminate' music and the need to 'let things be themselves'. While *2'33"* is clearly intended as a provocative piece, to challenge ideas of what could be called music, Cage experimented more widely with randomness in his compositions. One such work is his *Music of Changes* (1951), in which the musical elements were assembled based on questions asked of the *I Ching*, the *Book of Changes*. In that piece Cage goes beyond the approach of Mozart's dice game, and allows chance to

---

[117] The piece is in three movements. Various versions of the score exist; from these we learn that the tempo of the piece is 60 beats per minute, and that the piece can be played on any instrument. The instruction to the player for each movement is simply: *Tacet* (remain silent).

determine not only the choice of notes, but also their timing. Nonetheless, *Music of Changes* is still random only in composition, and not in performance.

Other kinds of indeterminate music involve randomness in different ways. A piece by Earle Brown called *Twenty-Five Pages* consists of 25 pages of sheet music, unbound, and capable of being read equally well when turned upside down. It is up to the performer or performers to choose how to make use of those pages.

Music incorporating random elements is sometimes termed *aleatoric*, from *alea*, the Latin word for a dice. Werner Meyer-Eppler, a Belgian acoustician, who coined the term in the early 1950s, explained that aleatoric music is 'determined in general but depends on chance in detail'. Through the twentieth century, serious composers experimented with introducing chance into their work, but few did it with the thoroughness of Iannis Xenakis.

Xenakis

stochastic (adj)

> 'Having a random probability distribution or pattern that may be analysed statistically but may not be predicted precisely'. (Modern, technical) 'Pertaining to conjecture'. (Earlier, 1660s) From

> *stokhastikos* (Greek) 'able to guess, conjecturing'. From

>> *stokhos* 'a guess, aim, fixed target, erected pillar for archers to shoot at'

Iannis Xenakis took the idea of music based on probability further than anyone before him, calling it stochastic music. Not for him Mozart's simple rolling of dice, or Cage's sorting of yarrow stalks. Xenakis had studied as a mathematician and an engineer, and was working at the dawn of the computer age, and that allowed him to use more sophisticated approaches to introducing chance into his compositions. He had been mentored by Olivier Messaien, and he was closely associated with the Musique Concrète movement, whose members built pieces by assembling fragmentary recordings of found sounds like creaking doors (we'd call them 'samples' today). The question of which sounds could be used to make music, and even what music actually is, was up for debate.

Metastaseis (Beyond the Static) is one of Xenakis's major early works. The piece starts with the string players playing a single note in unison. But immediately each player takes that note as a starting point for a minute-long *glissando*, sliding their finger along the instrument's fingerboard to create a sound of continuously varying pitch. Here's the big idea, though: each of the 46 players takes their *glissando* to a different pitch, so that the single note played in unison 'opens up' into a rich composite of many different notes. The end effect is a distribution of pitches: no longer a single note, nor either any recognized musical chord. It is a new thing, and to characterize it, it is necessary to talk about the statistical distribution of the choices made by the players. In discussions of Xenakis's work, the word 'density' comes up a lot, a word that can only describe a collection of notes and never a single note. In this music, the exact choice and placement of individual notes are not as important as the overall sound of the collection of notes, which has become a thing in itself, the important sonic entity. It's a musical example of the **individual–collective** duality.

That's just the start of Xenakis's explorations. He rejected the Musique Concrète approach of using found sounds and instead started to find new sounds by using the orchestra in new ways. The whole, if not always greater than the sum of its parts, is certainly something different, and must be understood by looking at the characteristics of the collection. A statistical description of a system, even when that system is a piece of music, is different from a description of each of its smallest parts.

In Xenakis's piece *Pithoprakta* (the name meaning something like 'actions by chance'), the pitches of the notes are in some passages based on a random walk, modelling the statistical mechanics of gases. Here is a portion of Xenakis's score for *Pithoprakta*.[118]

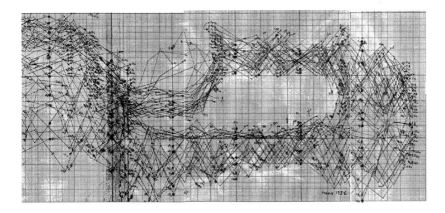

The horizontal axis of the chart represents time, the vertical axis represents pitch. Each of the many straight lines that make up the strange shape shown

---

[118] Image provided by kind permission of the Xenakis family: © Collection Famille Iannis Xenakis, DR.

represents a *pizzicato glissando* played by a single instrument (that's a plucked note, with the pitch being changed as the note decays). Xenakis has given general directions, but individual players must choose their exact path through this shape. Just as the eye, looking at his diagram, focuses on the overall shape rather than the individual line segments, so the ear hears the envelope of shifting sounds as a whole rather than picking out the trajectories of the individual notes. The collective sound (Xenakis calls these collectives 'sonic entities') counts for more than the individual notes. There is available on YouTube a performance of *Pithoprakta* that is accompanied by an animated graphical rendition of the score, created by Pierre Carré. Here are three images from various passages in the work.[119] The word 'texture' seems to describe both the visual images and the sound of the music itself equally well in the corresponding passages.

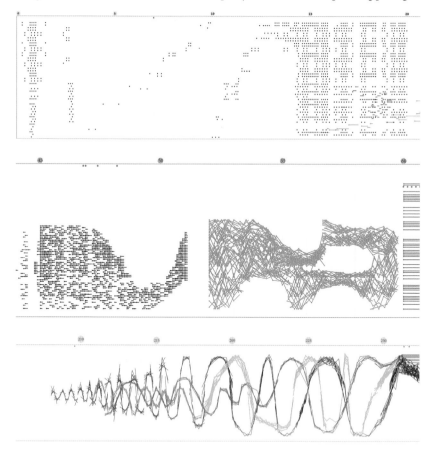

[119] Images provided by kind permission of Pierre Carré. Many thanks to M. Carré for providing these in higher resolution than can be captured from YouTube.

The first of these illustrates a passage early in the work: the dots represent a technique Xenakis denotes as *frappé col legno*, 'struck with the wood'. The effect on the ear is almost like hearing a Geiger counter or a hailstorm on a tin roof: a texture of sharp percussive sounds. The second image corresponds to the extract from Xenakis's score already shown, while the third image is later in the piece and is a series of glissandi: red lines are bowed, orange are plucked, purple are played *sul ponticello*—close to the bridge—while the green lines represent glissandi played with the wood of the bow *col legno*. As the image suggests, the music swoops and rises in contrasting waves. It is extraordinarily engaging to watch this animated score while listening to the piece, which can otherwise be challenging to hear without preparation or assistance.

Through his career, Xenakis explored the use of mathematics, and in particular probabilities, in many different ways. In his work *Analogique A* for nine stringed instruments, the music is made up of eight so-called screens. Each screen is a set of instructions that the players must follow for a set period of time, and each contains an array of cells, each stipulating 'a pitch region, a dynamic intensity and a density' for the notes to be played for the duration of the screen.

The flow of the pieces between screens is randomly directed, but uses specified probabilities for the transitions. These rules prevent the music from falling into complete anarchy. Xenakis regarded this piece and its companion *Analogique B* for electronic instruments as failures, because he felt that in performance, the audience were not able to discern the structure underlying the compositions. It remains, though, an extraordinary experiment in music-making.

Getting technical: Markov chains

In *Analogique A* Xenakis directs that the flow between musical units, or screens, is determined by numerical probabilities. For each of the eight screens he defined the probability that the succeeding screen would be each of the others. This makes a matrix of probabilities, one for each of the 64 combinations of 8 screens. So, for example, screen D is followed by screen H 21.6% of the time, while screen A is followed by itself just 2.1% of the time. Mathematically, this set of probabilities is called a transition matrix. When a transition matrix is applied iteratively, the resulting sequence of elements is known as a Markov Chain.

Markov Chains are used in many simulations where each new element is random. Unlike a series of coin flips, each event is not independent of what came before, but depends on the current situation. In a Markov Chain, each state of the system dictates the probabilities of the states that follow. In the musical context this gives more coherence to the piece, and keeps it from pure randomness.

Any artist who relies on chance must resist the tendency for their work to become mere chaos. But give the noise some structure, and randomness can become pleasant, even therapeutic. The sound of the waves breaking on the shore is unpredictable, but nonetheless has some structure. When we hear music, our brains listen out for structure, but are still open to surprise. We expect the composer's artistic sensibilities to steer a course for us between comfortable regularity and inspired creative leaps. Too predictable and the music is boring; too surprising and listening becomes unpleasant. In Xenakis's music, the mechanisms for balancing structure and randomness are unconventional and unfamiliar, but are still present, and they make for music that is challenging to listen to, but nonetheless fascinating and involving.

Generative music

Xenakis was part of the early experiments with electronic music: recordings on tape cut up and spliced and early techniques for synthesizing new sounds. Those pioneering sound production capabilities now seem rather primitive in the light of the capabilities of modern computers. Music of all kinds is now available in ways never imagined, and the facilities available for generating music automatically and randomly are now very impressive.[120]

In 1995 Brian Eno, founder member of the rock group Roxy Music and record producer extraordinaire,[121] was experimenting with a piece of software called SSEYO Koan. This was a system for automatically creating music in a structured but random way. The sounds produced by Koan were random sounds, but the software contained and constrained the random elements by allowing them to move within an overall structure. Eno dubbed this 'generative music'. Making such a piece of music requires the composer to define the overall shape of the music, much as Xenakis had done, but then relies on randomness to fill in the detail. When the piece is 'performed' (the computer program is run with the chosen parameters), the resulting sounds adhere to the style and shape constraints provided by the artist, while the random elements ensure that

[120] Not least, to support the need for background music in video games. The music must set the mood for the action, but must also be adaptive. It should be capable of adjusting to variable amounts of time spent in different phases of the game, and even responding to the intensity of the on-screen action.
[121] Eno also coined the name Ambient Music for a style of atmospheric, minimalist, but nonetheless evocative compositions. *Discreet Music* and *Ambient 1: Music for Airports* are two excellent Eno albums characteristic of this style.

each rendition is unique. Over the years, this Koan software has been refined and developed, and Eno has collaborated on a variety of different manifestations of these automatic music generators, including apps for smartphones and tablets. One of the current incarnations is called Wotja. For a modest fee, you download Wotja to your device of choice and listen to music created on the fly, never heard before, never to be heard again. The results are surprisingly listenable, though perhaps not very exciting, and so best used in a meditative, reflective context, or as background for working.

Not just music

I've taken music as my theme in this chapter to illustrate different ways in which randomness can be recruited into the creative process by musicians. But randomness is available to artists of other kinds as well. When Jackson Pollock poured, dribbled, and splashed paint onto a canvas on the floor, the resulting images were a blend of intention and chance. Some critics felt the random elements made the resulting work meaningless, raising questions of where meaning is to be found in abstract art of any kind, music included. Pollock himself noticed that people tried to find representational elements and more explicit meanings in the paintings, and so he broke away from giving names to the works, preferring numbers. Meaningful or simply random, the work is now highly valued. His *No 5, 1948* was reportedly sold in 2006 for $140 million.

Generative techniques have been applied to the visual arts as well. Artists define computer algorithms that take a simple set of rules and, incorporating randomness along the way, create images that are surprisingly powerful, visually satisfying, richly textured, and unique.

The Dada movement of the early twentieth century used random techniques for composing texts. The technique they called 'cut-up' involved taking existing texts and splicing them together in random ways, to create new material spontaneously. The beat writer William Burroughs experimented with cut-up approaches and wrote:

> The method is simple. Here is one way to do it. Take a page. Like this page. Now cut down the middle and cross the middle. You have four sections: 1 2 3 4…one two three four. Now rearrange the sections placing section four with section one and section two with section three. And you have a new page. Sometimes it says much the same thing. Sometimes something quite different.

David Bowie, too, has described how he used cut-up techniques in writing song lyrics, sometimes as inspiration, but sometimes directly: reportedly his *Moonage Daydream* was written this way. Which brings us right back around to music.

Breaking the pattern: Spontaneity

We've not yet finished talking about Brian Eno, a frequent collaborator with Bowie. *Oblique Strategies* is a work published in the 1970s attributed to him and painter Peter Schmidt. *Oblique Strategies* (subtitled *Over One Hundred Worthwhile Dilemmas*) takes the form of a black box containing a stack of cards, each card containing a gnomic suggestion to prompt a line of progress when you find yourself blocked in some creative activity. The original 1975 edition contained 113 cards and was put out as a limited edition of 500 copies. Speaking of *Oblique Strategies*, Eno has said:

> If you're in a panic, you tend to take the head-on approach because it seems to be the one that's going to yield the best results. Of course, that often isn't the case. It's just the most obvious and—apparently—reliable method.

You use the *Oblique Strategies* by drawing a card at random from the box, and taking this as a prompt for a lateral move in the project you're working on, a jump away from the head-on approach. Some examples of the prompts in the box:

- 'Remove specifics and convert to ambiguities'
- 'Disciplined self-indulgence'
- 'Turn it upside down'

and rather wonderfully:

- 'Honour thy error as a hidden intention'

(Which could almost stand as a motto for this section of this book. Errors are indispensable.)

Using *Oblique Strategies* is not so different from how Philip K. Dick used the *I Ching* in the writing of *The Man in the High Castle*. It's a prompt to inject a little surprise, an element of unpredictability that might make all the difference between a competent piece of craftsmanship and something truly thrilling.

As we shall see in the next chapter, it is not only in music and other art forms that randomness plays a central role in creating something new. Innovation in art is often about bringing together disparate elements in surprising and productive ways, and then exercising artistic judgement in selecting which variations to retain. Perhaps one of the most impressive capabilities of the human brain is our ability to imagine what does not yet exist. An artist can perform multiple experiments involving the random combination of elements in their head, based on (among other things) the stimuli around them and, imagining the results, they can select those which seem to have the most promise and bring them into physical existence. Randomness is at the heart of creativity; it's also central to the processes that have created all life on Earth: genetic transmission and evolution through natural selection.

Answer to:
Fortune favours the bold?

In this game, you need to find the sweet spot to stop. Being reckless and pushing your luck as far as you can is a bad idea. Imagine if you'd already flipped 11 cards and hadn't yet seen the jack. There'd be a stack of 11 dollar bills in front of you, and the turn of the next card would, with equal chances, either cost you all of them, or gain you one more. You wouldn't take that chance. On the other hand, being ultra-cautious isn't so smart either. Suppose you'd flipped just one card, and avoided the jack. On the next card, you're risking only one dollar on a 1 in 12 chance against gaining a dollar 11 times out of 12. Obviously, you flip that card.

Apply that same logic for each card in turn, and you'll find that if you come safely to the seventh card, you have a stack of six dollars. Now, on the next card there is a 1 in 7 chance of seeing the jack and so losing your six dollars, to set against a 6 out of 7 chance of gaining one more dollar. One in seven times (on average) you will lose six; 6 in 7 times, you will gain one. They are perfectly balanced. Arithmetically there is, on average, nothing to be gained or lost by flipping the seventh card.

Maybe you shouldn't go any further than that, though. Flipping the eighth card exposes you to a 1 in 6 chance of losing $7, against a 5 in 6 chance of gaining one. The odds are that that would be pushing your luck too far.

What are the chances of getting to the sweet spot of six or seven successes? Making six safe flips means that the jack is one of the last seven cards: the chances of that are 7 in 13, and you'd win $6. The amount you'd pick up would be on average $6 \times {}^7/_{13} = \$3.23$, for a net gain of 23 cents. What if you flip the seventh card? The chances of surviving seven flips is 6 in 13, and you'd win $7. Average amount: $7 \times {}^6/_{13} = \$3.23$. Exactly the same. So the game is in your favour, provided that you know when to stop, after six or seven cards.

CHAPTER 12

A Chance to Live

The two go hand in hand like a dance: chance flirts with necessity, randomness with determinism. To be sure, it is from this interchange that novelty and creativity arise in Nature, thereby yielding unique forms and novel structures.

Eric Chaisson, *Epic of Evolution: Seven Ages of the Cosmos*

Even unto the fortieth generation

You have two biological parents, four grandparents, eight great-grandparents, and sixteen great-great-grandparents. For every generation you go back, the number doubles. Roughly speaking, each human generation is around 25 years; so your great-great-grandparents' time (four generations back) was about a century before your time. How many of your ancestors lived 200 years ago? That would be eight generations, and so we double up eight times in total, to get 256 great-great-great-great-great-great-grandparents.

What about a thousand years? How many ancestors are there in your family tree if you trace it back that far? A thousand years would be 40 generations. If we continue doubling until we've done it 40 times, we get to 1,099,511,627,776. More than a trillion ancestors, a thousand years ago. But a thousand years ago there were fewer than half a billion people alive on Earth. How can that be?

Answer at the end of this chapter.

What are the Chances of That? How to think about uncertainty. Andrew C. A. Elliott,
Oxford University Press. © Andrew C. A. Elliott 2021. DOI: 10.1093/oso/9780198869023.003.0012

We're all individuals

In the film *Monty Python's Life of Brian*, Brian addresses the crowd: 'You're all individuals!' 'Yes!' they respond with one voice, 'We're all individuals!'[122] We laugh at the paradox, but Brian is right. Despite their unison response, the members of the crowd are all individuals. Even identical twins, sharing a common genetic inheritance, the same DNA, will develop differently, not only due to the differences in their experiences, small or large, but because their DNA, although the same at the start of their lives, is susceptible to damage over their lifetimes. Our identities are the consequence of random, unpredictable processes, and this makes diversity an essential part of life on Earth. If there is life beyond our planet, it's a fair bet that the development of those creatures is based on similar principles of inherited characteristics, and will have arisen from the same interplay of variation and selection.

Our genetic identity arises from an exquisite balance between determinism (almost everything in our genetic make-up is owed to our parents) and randomness (our individual DNA is a shuffled combination of those contributions from two parents). That balance in the genetic code plays out as a balance between stability and novelty in the living creature, extended over generations, driving incremental change.

It's beyond the scope of this book (and my expertise) to describe the topics of evolution and genetics in any extensive way. But chance is deeply involved in the processes whereby life proceeds from generation to generation, and I'll try to sketch out just a few aspects that show how chance becomes involved (and consider some probabilities along the way). After all, it's one of the big questions: what were the chances that life on Earth would evolve in the way it did?

Genetic chances at the start of life

Every cell of your body contains genetic material. This is organized into chromosomes, strands of DNA containing the 'genetic code'. These molecules, bunched-together strings, are interpreted by mechanisms in your body's cells as instructions for building and maintaining your organs and systems by making proteins. Most of our cells are somatic ('bodily') ones, and each of these cells contains two versions of every chromosome, one version from each of your

---

[122] And then one of them pipes up in a timid voice: 'I'm not.'

parents. So each cell contains a full copy of the DNA donated by your mother and that donated by your father, brought together at the moment of your conception—46 chromosomes in total, usually. You carry this genetic material with you all through your life, in every cell of your body. The chromosomes from father and mother pair up, all apart from the twenty-third, where the chromosomes are similar in the case of genetic females (and designated XX), but different in the case of genetic males (XY). It has been one of the most surprising discoveries of modern science to learn that this genetic material acts as a digital code. Just as a computer program is brought to life and meaning when properly interpreted by a compatible computer, so the genetic code, when properly interpreted by the cells of the body, reveals its function. But the machinery that processes this code is chemical in nature, and the physical expression of that genetic material occurs through messy, wet, biological processes. Still, the instructions are essentially digital—coding[123] for the creation of proteins. And because this is at root a digital, logical process, it can in some ways be analysed successfully with mathematical techniques.

While the somatic cells each carry 46 chromosomes arranged in pairs, the cells used to transmit life—the gametes (sperm for males, eggs for females)—have only one of each chromosome, 23 in number. Most cells replicate by making complete copies of themselves, but the 'germ cells', located in the ovaries and testicles and responsible for producing the gametes, are different. Through a process called meiosis, they create the gametes, which contain incomplete copies of the genome, each bearing only half the organism's genetic material. Which half? Well, that's where chance comes in.

Each gamete ends up carrying a random selection (on average, 50 per cent) of the genetic material of the organism. Meiosis, the process of creating a sperm or an egg, starts when the chromosomes in the germ cell make copies of themselves, resulting in four of each. Then, there begins a process of shuffling (*recombination*) whereby pairs of chromosomes (one from each parent) 'cross over' and exchange stretches of their DNA in a random fashion. There are certain locations in the DNA where crossing over is more or less likely to occur, but this is nonetheless a random process. It's as if you had two parallel translations of the same text, and you chose to read long passages from one version, but occasionally you randomly switched over to the other. The resultant text would have come entirely and only from the two alternative versions, and would probably make sense, but the exact way in which the texts were spliced

---

[123] Sometimes coded in extraordinarily complex ways.

together would be random. In all probability, every reading you made using this process would be different.

How random is this recombination? What are the chances that one sperm cell's DNA content matches that of another from the same individual? It's been estimated that around 26 splices occur when DNA is recombined in sperm production, and around 45 splices happen when an egg is produced. If those splices occurred always at the same locations, that would mean 2^{26} (around 67 million) possibilities for sperm, and 2^{45} (around 35 trillion) possibilities for eggs.

But the splices are not always made in the same locations: so the calculation of possibilities is a lot more complex (and the numbers get much bigger) if you take into account that recombination can occur at different locations in the genome. One estimate of the possible number of variations for a single chromosome arising from typical levels of recombination is around 10,000. Considering the recombination process alone, this means that for 22 chromosomes[124] the number of unique combinations is then around 10^{88}, a number which is far, far bigger than the number of humans that have ever lived.

If you consider that the number of sperm produced by a male over a lifetime is of the order of a trillion (10^{12}), it is then virtually certain that no two of those sperm will carry the same genetic material. Going further, since the number of people who have ever lived is of the order of 100 billion, it is fair to say that the chance that two human sperm cells have ever contained identical genetic material is still so vanishingly small as to be a practical impossibility.[125] For women, each of whom may produce a million eggs (very few of which actually go through the ovulation process), the probabilities of producing identical genetic material in two eggs are even slimmer.

But recombination of chromosomes is not the only source of randomness: after the recombination process, the germ cell contains four sets of 23 chromosomes, which are then split into four gametes, cells with a single set of chromosomes each. The genetic make-up of each of these gametes involves further random choice, as 23 chromosomes are chosen, each from four possibilities that are present in the cell at that point. How many possibilities arise simply from making this four-way choice 23 times? Seventy trillion.

So mother has the egg, containing a randomly determined subset of her full genome, father has the sperm, likewise containing (on average) 50 per cent of his genetic inheritance. Through a final, little understood (but much investigated)

---

[124] The X and Y chromosomes are not recombined in this way, as they do not match.
[125] 1 in 10^{65}.

process of random selection involving social encounter, flirtation, courtship, and passion, boy meets girl, and in a way that is far from pure chance, and yet still far from deterministic, the germ cells are given the opportunity to be brought together.

In that final random process, sexual intercourse, just one of the male sperm cells wins out over hundreds of millions of rivals, in the race to fuse with the female egg cell. When the two germ cells fuse, the sperm cell's 23 chromosomes join with the egg's 23 to create the full complement of 46. The cycle of life begins again, with a new and novel set of DNA instructions for the creation of a new individual. For that new person, even though every component of their genetic code comes from their mother or father, their particular mix of genetic material will be unique in the history of humanity.[126]

Boy or girl?

By convention, the 23 pairs of human chromosomes are numbered 1 to 22, with the twenty-third chromosome designated as X or Y. Biological sex is determined by this twenty-third chromosome, and specifically by the differences between X and Y. In most human females, the twenty-third pair consists of two X chromosomes. This means that all egg cells will contain an X chromosome.

Human males have, as the twenty-third pair of chromosomes, one X and one Y. When a male germ cell produces sperm cells, half will contain an X and half will contain a Y chromosome. When an egg is fertilized with a sperm carrying an X chromosome, the baby will end up with two X chromosomes (XX), and so will be genetically female. If the sperm carries a Y chromosome, the baby will be XY and will be genetically male.

That's the general pattern, but life isn't always so cut and dried. Recall that after recombination, the human germ cell has four copies of the genetic material. Usually the result is four gametes (sperm in males, or eggs in females), each with one complete set of DNA. But this logical process is carried out by imperfect biological mechanisms, and that introduces scope for variation. So the allocation of chromosomes to gametes can sometimes happen in different ways. In an occasional case the sperm might carry both X and Y chromosomes, or an egg might end up with XX, rather than just one X. The result, after fertilization, might be a genome with 47 chromosomes, including two Xs and a Y. That

---

[126] It sometimes happens that the fertilized egg will split into two, each containing a full copy of the combined genetic material. This is how identical twins are formed.

results in a condition known as Klinefelter syndrome, a variety of intersex.[127] What are the chances of this happening? Around 1 in 500 live male births are XXY intersex.

Triple-X genomes can also occur when the sperm contributes an X to an egg that already has two X chromosomes. This is rarely diagnosed, as the symptoms are not easily detected. Turner syndrome is where the fertilized egg carries only one sex chromosome, a single X, and this occurs in between 1 in 2,000 and 1 in 5,000 live female births.

Sexual reproduction evolved approximately 1.2 billion years ago. Earlier organisms cloned themselves—and simple organisms still reproduce in this way. When bacteria make copies of themselves, opportunities for variation occur only through the making of imperfect copies that are flawed but nonetheless viable. But creatures that reproduce sexually have many more opportunities to exploit the potential not only for mutation, but for different combinations of genes. By shuffling together the DNA from two healthy individual organisms, each generation of offspring carries DNA that has never existed before in that precise form, but which is very likely to create viable bodies.

Covid-19 is a disease caused by the virus SARS-CoV-2. The genome of this virus is a single strand of RNA, approximately 30,000 bases[128] long. This genome codes for four structural proteins that make up the virus particle, the viroid. Three of these make up the outer structure, the fourth holds the RNA genome.

Viruses rely on the host cells to replicate, hijacking the host cells' copying machinery. While DNA replication includes an error-correcting mechanism, RNA replication does not, and this makes RNA much more prone to transcription errors that might then persist into successive generations. In practice, this means that RNA viruses are less stable and more likely to mutate randomly. If two similar viruses are present in the same cell, then there is a possibility for stretches of RNA to cross over between the two viruses and, through the operation of chance, produce a new version that is better adapted to more efficient infection of the host species. The SARS-CoV-2 virus is very similar to one found in bat populations, which suggests that it might have originated in bats, though it is currently thought possible that the transmission from bat to human may not have been direct, but may have involved pangolins.

---

[127] A general term for a variety of sex development disorders.

[128] Bases, not base pairs. RNA is single stranded, unlike the double helix of DNA. Some viruses (such as smallpox) have a DNA genome, but coronaviruses are based on RNA.

Mutation

A fertilized egg is only a starting point. The twisted, tangled strands of DNA are a code, and that code is read and interpreted in chunks: it is like a computer language, but not one that has been purposefully designed. The language of DNA has evolved to produce proteins that then build the human (or other) body: the translation of genotype (the genetic instruction set) to phenotype (the result: the observable characteristics of the body).

The underlying digital nature of these instructions accounts for the way in which genetic material is combined: it is a shuffling rather than a blending. No amount of shuffling of a pack of cards would give a suit intermediate between hearts and spades: in the same way, the reproduction process chooses between, but does not blend, genes.

Some processes, though, do cause actual change to the genes. These are mutations. The growth of all living organisms relies on a process of copying, and as our bodies develop from the first fertilized eggs, each new cell that is made contains a copy of our own DNA. If the copying process were not highly accurate and reliable, the DNA would quickly become incoherent and the organism would fail.

The copying ('transcription') of a DNA strand is followed by an error-correcting stage, which checks the copy against the original. It has been estimated that each of these processes has a failure rate of around 1 in 100,000. For an error to slip through both processes we must multiply together the chances of each failure, giving, overall, a 1 in 10 billion error rate per base pair copied. Since human DNA has around 6 billion base pairs, this results on average in 6 mutations for every 10 DNA transcriptions. And since every cell in your body (and there are around 40 trillion of them) contains a copy of your DNA, more than half of them will contain a mutation, an error arising from faulty transcription.

Transcription errors are not the only cause of mutation. DNA is a complex chemical molecule. It is exposed to damage in many ways; radiation is one of these. Being bitten by a radioactive spider may not give you spider powers, but radiation can and does damage DNA, and can cause mutation. Sunlight is radiation, too, and can also cause damage to the DNA in skin cells, and that can result in skin cancer. Chemicals can change DNA; compounds associated with smoking can cause damage to cells in the lungs and other organs—cancers of the throat, the liver, the kidneys, the bladder can all be triggered as the poisons from cigarette smoke move through the body.

Our DNA is vulnerable to assault every day, but there are also defences. Our cells have evolved mechanisms that can sometimes repair DNA. Of all the incidents that damage our DNA, only a very few result in lasting harm, but over a long life, the damage accumulates—which is why, generally speaking, cancer is an old person's disease. In most cases, mutations will be harmless; nonetheless, the greater the exposure, the greater the risk of harm. Smoking one cigarette will not kill you. But each cigarette you smoke has a chance of doing damage, and in the long run, heavy smokers run a very high risk of seriously damaging their health.

In a very small number of cases, the mutations to DNA may even be beneficial, though the superpowers they confer may be hard to detect. And only when mutation affects the germ cells (which produce eggs in females, sperm in males) is the mutation liable to be passed to the next generation. And those infrequent variations are what drives evolution.

Fruit machine

Let's step back into the casino. Which game is more played than any other? The one-armed bandit, the slot machine, the fruit machine. Three wheels roll before your eyes and then come to a halt. If they end up in one of numerous favourable configurations, there may be a payout due to you. The coins clatter out of the chute, the level of noise deliberate so that your neighbouring players hear evidence of your good luck, and are encouraged in their efforts to beat the odds.

The typical slot machine has perhaps 20 symbols on each of three reels. This makes for 8,000 possible combinations (multiply the options for each of the three independent reels: $20 \times 20 \times 20$).[129] Many fruit machines have options for the players to take further action—for example, to roll again, perhaps freezing one or more wheels, thereby improving the odds of winning, and giving the punter a semblance of control over the process. Now imagine a variation of the typical slot machine. This one has 10 reels, each with 10 symbols on it. For each reel, each symbol has an equal chance of showing in the winning frame. The jackpot is paid out when each of the 10 reels is in the right place. What are the chances of winning on a single spin? By now, you know how these calculations are done. Each wheel is independent, and each has a 1 in 10 chance of coming

---

[129] You'd think this made for a 1 in 8,000 chance of winning the jackpot main prize, but not so. Modern slot machines are computer controlled, and the probabilities of winning or losing can be configured programmatically; so the simple rules of probability don't necessarily apply.

to rest at the necessary spot. Multiply those chances for each of 10 wheels, and you end up with 1 in 10^{10}, or 1 in 10 billion. On average, you would need 10 billion spins before the wheels ended in the winning positions.

But suppose that you could freeze any of those wheels that had indeed ended up in the right spot, and only re-spin the remaining ones. How does that change the odds? How many spins would be needed to get the winning 1 in 10 billion combination?

I coded up a simulation of the situation to see how many spins of that fruit machine would be needed to get to the winning combinations. Each time, my computer spun the 10 reels, froze any reels that were in jackpot positions, and re-spun the remaining unfrozen reels. How many spins would it take to get all 10 reels into the right position? Here are the 1,000 results. The position of each dot against the y-axis represents how many spins it took to get all 10 reels in the right place and so win the jackpot.

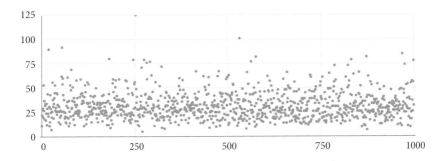

In my simulation, the smallest number of spins needed was 5, the biggest was 125, and the average was just under 32. That's many orders of magnitude away from the number of spins needed for the machine without the freeze function, where it would take on average 10 billion spins to hit the jackpot.

Why this diversion on fruit machines? Thinking about chance never comes easily, but when it comes to very small chances over very many repetitions, our intuitions are more or less useless. How could the 520 million base pairs that make up a banana's genome have come together by pure chance? The odds of that chance occurrence would seem to be inconceivably small. But if we think about evolution as a kind of fruit machine, where all the base pairs in the genome need to line up in the right way, then it's the kind of fruit machine that has a way of freezing wheels. Evolution has a way of remembering: it's the DNA that records and transmits information from one generation to the next. Each generation, by and large, retains any advantages held by previous generations.

Only a small proportion of the reels are spun each time: the benefits of good spins get preserved.

This is of course an outrageous oversimplification. There is no single jackpot for a banana's DNA, and the freeze mechanism doesn't fully 'lock in' desirable characteristics. It's not just a matter of getting the right symbols lined up against each other; there are complex interactions between parts of the genetic code. But the fundamental principle is the same. Each generation does not need to start from scratch. Each has the advantage of the billions of years of trial and error accumulated in the DNA of its ancestor generations. (Oh, and don't play the fruit machines. The odds are awful.)

Evolution

Life on Earth started some four billion years ago. From our Last Universal Common Ancestor (about whom more later), right up to you, dear reader, and me, there is an unbroken chain of heritage. Our ancestors include many weird and wonderful creatures, but every one of them managed to survive the perils of life long enough to reproduce, and each benefited from the adaptation of its ancestors to the environments in which they found themselves. Through 'snowball Earth', through warming oceans, through the consequences of asteroid impact, our genetic material was shaped by random variation and evolutionary selection. It is a profound irony that while in the short run errors in the accurate propagation of DNA are deleterious, in the long run those errors are essential to adaptation and hence survival. Randomness is what makes it all possible.

The ones who got lucky

Many people find it difficult to accept natural selection as the principle behind the wonderful variety and beauty of the plants and animals that populate our world, not to mention the marvel that is the human brain. How could such richness, such variety, such excellence have arisen purely by accident? Is it truly possible to believe that ecosystems could have been shaped simply by random changes to DNA? Is it, as Fred Hoyle said, as if a hurricane tore through a junkyard and assembled a Boeing 747? Richard Dawkins explains, in his book *Climbing Mount Improbable*, one way of resolving this seeming impossibility. In his analogy, Mount Improbable has one face that consists entirely of almost vertical cliffs. You do not even think about scaling those cliffs directly. By analogy, we do not think about the improbability that viable DNA for a complex organism could have been assembled by chance all in one go. But there is another route up Mount Improbable. Go round the back, and there is the gentlest of slopes, although it's a very long path. Step by step, over a very long time, the DNA required for complex organisms can be assembled, mutation by mutation, recombination by recombination. Mount Impossible is so high we cannot easily imagine that there could be a gentle route to the summit. But there is, although it is also very long indeed.

In thinking through this question of the improbability of assembling three billion base pairs into a viable genome, we bump up against the dualities of thinking about chance.

The question of the improbability or otherwise of evolution gives us a powerful illustration of the **individual–collective** duality, and the difficulty we have in thinking about large numbers. Our thinking is shaped by our experience of events over time spans consistent with our own lives. But with evolution we are dealing with time scales and population sizes that are way beyond what we as humans experience, and to which our intuitions are not well suited. A single lifespan does not equip us to make good judgements about what can happen in four billion years. All human history (since humans could write) covers a mere 5,000 years. That's about 200 human generations. But that span of time represents not much more than one millionth—one part in a million—of the time that life has existed on Earth.

Then, since we are looking backwards but thinking about forward probabilities, it's also an example of **foresight–hindsight**. Certainly the evolutionary path that led to us being here now was improbable. Perhaps the biggest, most difficult-to-believe coincidence of all is that we are here, now, and that we happen to be here now simply by chance. But it is only because we are here now that we can even ask the question. Like me, you are the product of a one-in-a-gazillion sperm fusing with a one-in-a-bajillion egg. We are both entirely improbable, you might say nearly impossible creatures. But here we are, you and I: me writing, you reading. Because we are the ones who got lucky. I am me because that one sperm of my father's fused with that one egg of my mother's. He became who he was in the same way, and so did she. And the same is true of his parents and her parents before them. We are all improbable to an unimaginable degree. And unlikely as we all are, we seldom think of the even more unimaginable children, the people who never were, the possible fusions of sperm and egg that never happened. We are the ones who got lucky in more ways than we can possibly count. And we look back, from our immensely privileged place on our family trees and think we are special. We are not. We are only lucky. It's an extreme case of survivorship bias.

Randomness–meaning: why are we here now? It is tempting, and understandably so, to regard our species, *Homo sapiens*, as some sort of end goal of evolution. But we are simply one branch on a very untidy tree. We have some remarkable abilities, but we are far from being perfect creatures. Other species, other products of evolution, can claim superiority over us. We cannot fly. We cannot swim very well. Our eyesight and hearing are relatively poor, and our sense of smell is rudimentary. We are not very strong, physically. Ants vastly outnumber us. We can claim superiority in the brain department, although Cephalopods, Cetaceans, and other animals also have strong claims for high intelligence, even if their types of intelligence feel a little alien to us.

Let's not forget the other two dualities. Clearly, when it comes to evolution, we can see the interplay of the **disruptive** and the **opportunistic** aspects of chance, as the tiny proportion of happy genetic accidents and crossovers accumulate to produce creatures that are on the whole better adapted with each successive generation. And if there was no variation in the genetic code from individual to individual, there would be no potential for selection of favourable traits. What is important for evolution is the balance between stability and experiment, between the **uniform** and the **variable**.

My family and other animals

If DNA is an instruction book for the making of an organism, it can also be read as a history book. Through comparing stretches of DNA, geneticists are able to identify with a very high degree of confidence how much genetic heritage is shared between organisms. And because of advances made in recovering ancient DNA, they are also able to measure the relatedness to bones of earlier humans, even those of other species, such as Neanderthals. Much of this work depends on an understanding of probability and statistics. The changes in the DNA when it is transmitted from a pair of parents to a child over one generation may be quite random, but that randomness can be modelled and the probability distribution understood. The law of large numbers lets us make strong inferences about the rate of change of DNA over many generations. In this way, the evolutionary process can be quantified and measured. This allows the history written in our genes to be read with increasing confidence.

The range of organisms on Earth is frequently portrayed as a branching tree. But geneticists are discovering that it is a very odd tree indeed; mostly it follows a familiar branching pattern, but sometimes branches come together and merge. There is now strong evidence that modern humans and Neanderthals, whose branches separated around a quarter of a million years ago, were once again interbreeding in much more recent times. Around 50,000 years ago the branches touched and fused: we all carry some Neanderthal DNA.

It's thought that near the base of the evolutionary tree it may look more like a network. We can trace our descent to earlier and earlier forms that share common DNA with more and more of the species now on Earth. Take that to the limit, and we reach a place where all those forms come together, from a single origin. Where the branches meet is an organism that has been called the Last Universal Common Ancestor (LUCA), calculated to have lived roughly four billion years ago. Every organism on Earth has this ancestor in its heritage. The LUCA is probably not the very first living creature, but it is the one whose descendants survived to populate the Earth.

If four billion years of evolution can incorporate all the changes needed to link a single-celled LUCA to you and me, how many generations does that represent? That is a very hard question to answer, bearing in mind that for bacteria such as *E. coli* a generation is around 20 minutes, while for humans a generation is around 25 years. The balance between those extremes must lie with the earlier, simpler, organisms with very short generations, but I can find no credible estimate of the total number of generations for the bulk of the time between LUCA and ourselves.

We can be more certain of the pace of evolution in times nearer to the present. In his book *The Ancestor's Tale*, Richard Dawkins gives the following estimates of time and generations to reach the Last Common Ancestor (LCA) relating to more recent branchings of that tree.

> The LCA for all monkeys and apes: 40 million years ago (3 million generations)
> The LCA for all mammals: 180 million years ago (120 million generations)
> The LCA for all chordates: 560 million years ago (270 million generations)

Many mutations neither help nor harm: they are neutral. This means they are selected neither for nor against in the evolutionary process, and so they can be used as an evolutionary clock. Neutral mutations accumulate at a steady rate over time, so even though the occurrence of mutations may be completely random, the law of large numbers means that, given enough generations, the average rate of change is predictable. This means that the difference between comparable stretches of DNA from different organisms is proportionate to the time since those organisms diverged from a common ancestor. For example, it has been reckoned that bird DNA diverges at the rate of around 2 per cent per million years, and that metric is invaluable in understanding that part of the evolutionary tree. It's a remarkable example of how statistical regularity in a random process can provide a reliable and useful measurement, even though it arises from mere chance.

Answer to: Even unto the fortieth generation

A thousand years; 40 generations. With every generation, there is a doubling of the number of ancestors. Double up 40 times, and you reach a trillion ancestors in your family tree. And yet, a thousand years ago there were not even half a billion people alive. How could that be?

The explanation, of course, is that while there may be a trillion roles to fill for your 38-times great-grandparents, those roles are not filled by a trillion different people. Imagine your family tree branching over and over again. Go far enough and sooner or later you'll find the same person on two different branches of that tree. Go further and you will find, more and more, the same people filling multiple positions. Eventually some people will be filling hundreds, even thousands of roles on that exponentially branching tree. It must be so: if there are a trillion roles to fill, and at most 500 million people to fill them, it must mean that, if you go back 40 generations, a thousand years, then on average each person then alive must appear in your family tree 2,000 times. And that means that we are all related, many times over. Hi, cousin!

CHAPTER 13

Random Technology

It is remarkable that a science which began with the consideration of games of chance should have become the most important object of human knowledge.

Pierre-Simon Laplace

Noisy colours

'White noise' is well known as a hiss that is effective at concealing distracting noise in your environment. But do you know the meaning of these other 'colours of noise'?

- Pink noise
- Blue noise
- Green noise
- Brown noise
- Black noise.

Answers at the end of the chapter.

Breaking step

Organized forces can be dangerous. Traditionally, soldiers marching in formation are ordered to break step when crossing bridges. The reason? Synchronized footfalls could set up resonances in the structure of the bridge, causing a build-up

What are the Chances of That? How to think about uncertainty. Andrew C. A. Elliott,
Oxford University Press. © Andrew C. A. Elliott 2021. DOI: 10.1093/oso/9780198869023.003.0013

of stress and possible failure. The Tacoma Narrows Bridge collapse in 1940 shows what can happen to a bridge when it starts to resonate (although in this case the resonance was caused by the wind). It's easy to find footage online showing the bridge shaking itself apart in a very dramatic fashion. The Millennium Footbridge across the Thames in London became known as the 'wobbly bridge' when it first opened in 2000: the behaviour of pedestrians and their tendency to synchronize their footsteps caused the bridge to wobble perceptibly and alarmingly. It was closed within days and had to be fitted with special damping mechanisms to break the patterns established by the footfall of those walking across it. Too much uniformity can be a problem.

Computer systems, too, can be vulnerable to the effects of synchronized, uniform behaviour. Any time you connect to a website, the server computers that respond to your request may need to read data from databases, write to files, communicate through network links, and perform a thousand other tasks, many of which will involve sharing resources. If a computer receives many similar requests at the same time, the resulting spike in demand can overwhelm its resources and lead to sluggish response or system failure. If the congestion were to be handled naively, say by retrying each unsatisfied request after a set interval, the congestion would recur repeatedly, and even be amplified in a kind of resonance effect. The solution? To delay requests by a *random* time interval to break the lockstep behaviour and inject a little variability.

Modern technology and computers in particular might seem to represent the ultimate in predictable behaviour. Computer programs are written to behave in expected ways, and to follow their instructions, invariably. But most computer systems need at some point to connect with the messy world outside their own bubble. The pattern of how and when users connect to websites and other services is unpredictable—so to test the behaviour of a system under more realistic conditions, programmers must simulate erratic patterns of access, and chance failures. Part of this testing involves the equivalent of breaking step: staggering the load in random ways to avoid artificial spikes. Put another way, the programmers must mix in a little noise.

Beautiful noise

The roots of the word **noise** are unclear. Etymonline.com suggests:

- Old French *noise* 'din, disturbance, uproar, brawl' and goes on:

○ According to some, it is from Latin *nausea* 'disgust, annoyance, discomfort', literally 'seasickness'

○ According to others, it is from Latin *noxia* 'hurting, injury, damage'.

Whichever derivation is right, the prevalent sense is negative, unpleasant, unwelcome. There is a modern technical meaning as well, also with negative connotations:

- Irregular fluctuations that accompany a transmitted electrical signal but are not part of it and tend to obscure it.
- Random fluctuations that obscure or do not contain meaningful data or other information.

But noise is not always unwelcome: sometimes a little chaos can mask an unwanted signal. White noise is a combination of all audible frequencies of sound,[130] with equal weighting at all frequencies. The result is a waveform that is random in structure, and that is perceived as a continuous hiss. In a busy environment, this has the effect of suppressing other ambient sounds, so that half-heard conversations and other noises are muffled and become less distracting. Frosted glass does the same job for the sense of sight: a frosted glass panel obscures the detail of what is on the other side, preserving privacy and reducing distraction, while still allowing light through.

We recognize different patterns of randomness as different distributions. And noise, too, comes in many different forms, characterized by terms borrowed from the theory of colour. White noise is a random fluctuation in an audio signal, a mixture of frequencies where all are equally weighted. But white is not the only possible colour. If the frequencies are mixed in different proportions, the result will be a different 'colour' of noise—a hiss that sounds different to the ear. So pink noise is weighted towards lower frequencies. This makes an audible difference: some feel that pink noise is more effective than white noise as a mask for ambient sounds.

'Red noise' is to pink noise, as pink is to white: that is, it is even more biased towards low frequencies. By contrast, 'blue noise' is biased to high frequencies, while purple noise is even more biased in that direction. These colours of noise are not just fanciful conceits. They are useful and precise characterizations of the patterns of randomization occurring in physical systems. Because randomness can be analysed, characterized, and understood, a knowledge of the statistical

---

[130] White noise is named by analogy to white light, which is a mix of all colours of the visible spectrum.

characteristics of noise allows engineers to manage it: sometimes with the aim of defeating the noise, sometimes aiming to exploit it.

Card players insist that the deck be well shuffled before dealing, eliminating any trace of order or pattern arising from the previous game's play. Traffic planners in cities arrange the timing of traffic lights to patterns of build-ups at sensitive road junctions. Soldiers break step to cross a bridge. All these involve the injection of an element of randomness into the system, to ensure even distributions and avoid patterns. All involve introducing noise to eliminate an unwanted signal.

Jittering

A teacher wanting to teach a class something about data exploration might do a class survey asking how many children are in each child's family and how many pets each family has, plotting the data in a scatter plot like that shown in the first chart here. Because the number of children and the number of pets are both whole numbers, there is a problem. The teacher has collected 50 data points, but there are only 15 unique combinations of number of children and number of pets; so we see only 15 dots on the chart. This fails to convey any sort of visual sense of the weight of the data that's been collected. The problem is too much regularity.

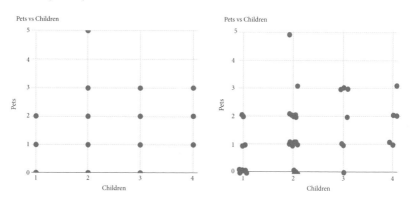

If, however, we cheat a little, and adjust our whole numbers by adding a little random adjustment to each coordinate, then the more popular data points become clusters. So, the point that represents two children and one pet becomes a blob rather than a simple circle, showing that many data points fall there. This technique is called jittering—adding in a little uncertainty to stop too much regimentation. Jittering of various kinds is widely used in computer systems to avoid over-regularity, and to make variability more obvious.

Dithering

I remember, as a child, squinting more closely than I ought to have at colour pictures in weekly magazines,[131] and marvelling at how the many colours in each picture were formed from just four colours of ink. Looking closely, I would also see that the dots from which each image was formed were arranged in a pattern of rosettes. The rosettes arise as a form of moiré pattern, which is formed when two or more regular patterns interact. In four-colour printing it is usual for each of the colours to be printed as a half-tone screen, an array of dots arranged in equally spaced rows and columns, but of variable diameter. Usually, the screens for cyan, magenta, and black are angled at 30 degrees to one another, which, for three screens, makes the moiré rosettes as small as possible.[132]

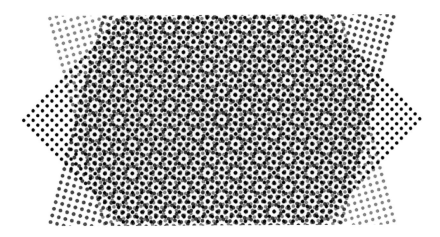

It's the regularity of these screens that makes the rosettes particularly noticeable. The human brain is very ready to notice patterns, and that makes these artefacts annoying. The grid patterns of the screens are an unwanted structure that imposes itself and becomes an unwanted signal. What's needed is a little less **uniformity**, a little more **variability**, and that can be provided in the form

---

[131] I say weekly magazines. I might mean comic books.
[132] Yellow is handled differently in traditional half-tone printing because of the lightness of the colour.

of noise. This approach is called stochastic screening and involves computer-generated screen patterns that make use of blue and green noise.

In the same way that blue light is biased towards high frequencies of the visible spectrum, so blue noise has more power in higher frequencies. When this statistical idea of noise is applied to visual patterns such as the spacing of dots on a screen pattern for printing, the frequencies correspond to the spacing of the dots. Here, high frequencies mean more weight for closely spaced dots. Equally sized dots placed randomly, based on the statistical parameters of blue noise, form a patternless half-tone scale.[133] Through clever use of randomness, an unwanted signal (the moiré effect of regular patterns) is suppressed. This use of randomized screen patterns is called dithering.

Modern inkjet printers use four colours of ink: cyan, yellow, magenta, and black. By combining these with dithered placement of dots, a very wide range of colours can be produced. Below, the image on the right is from a photograph of a printed copy of the image on the left. The use of dithering by the inkjet printer to render the subtle gradations of colour can easily be seen.

Scrambling

Shortly before the Second World War, Bell Laboratories developed a technique for adding and subtracting audio signals to and from one another. They realized that this could be used for secure transmission of audio messages: voice recordings or real-time telephone conversations. A recording was made of random white noise, and this was used to create two identical

---

[133] This is like Xenakis asking performers to play randomly, but to adhere to his statistical directives.

vinyl record discs. To disguise an audio signal, the message-carrying signal was added to the noise recording. When mixed together, the resultant audio was unintelligible and would be of no use to anyone listening in: it was scrambled. The receiver of the message, though, in possession of a matching recording of the noise used for scrambling, could use Bell's technology to subtract the noise, and in this way reveal the original message. While this scheme was both ingenious and effective when properly set up, it was rather impractical. As you might imagine, the proper set-up and synchronization of the two noise sources was very difficult, and the technology was quickly replaced as electronic and computer technologies developed through the twentieth century.

Randomization, though, remains essential to secure communications. The coding used to encrypt and decrypt secure messages passed over the internet and through other channels depends on keys generated and shared between the message-maker and the message-receiver, and the generation of these keys depends on random numbers. Any weakness in the methods becomes a point of vulnerability that can be exploited by a potential eavesdropper who might intercept and decode the message.

Searching the solution space

The power of modern computer systems means that many problems in engineering and other disciplines that in the past have been tackled by analytical methods—say, by working out a formula—can now be done using numerical methods. Hard questions that may have been impossible to answer in practical ways can now be tackled using computer algorithms, and many of the algorithms rely on random numbers.

Whether you are designing an airplane wing, or a queueing system for a supermarket, finding the optimal solution will often involve finding the best values for a set of numerical parameters (the width of the wing, the number of till operators). With enough computer power at your disposal, one approach is simply to try out different parameters over and over again, and look for which sets work the best. This is called exploring the solution space, and rather than doing this in a systematic way, which may be very time consuming and slow to reach a solution, it may be quicker to search the solution space in a random way. Some of these approaches mimic the process of evolution: random variation, selection for fitness, and preservation of progress through inheritance.

Simulating evolution

In the 1980s, biologist Richard Dawkins created a suite of computer programs called the 'Blind Watchmaker', designed to illustrate just how effective the combination of random mutation, directed selection, and inheritance can be at achieving gradual change. This is ancient history in terms of using computers to simulate evolutionary processes, but Dawkins's code is still available, still runs, and still provides a clear and vivid demonstration of the fundamental principles of evolution: random variation, inheritance, and selection.

When it's fired up, the Blind Watchmaker program shows an initial display of a 'biomorph'. This is a primitive computer graphic, its shape dictated by the interpretation of a small simulated genome—simply a collection of numbers that determine what shape is to be drawn. This 'ancestor' biomorph is displayed centrally on the computer screen, and it is surrounded by a number of mutant 'descendant' variations formed by tweaking the numbers that make up the ancestor's genome. This simulates inheritance with random variation.

The user of the program is responsible for the next stage: selection. You choose which of the newly created generation of offspring should survive, and carry forward the line of descent. Your selected offspring then becomes the graphic at the centre of the display, and is in turn surrounded by its own offspring. And so you can continue, generation by generation. If you have an idea in mind of what sort of shape you are 'aiming for', this mutation/selection process can quite rapidly offer up something that fits the goal.

I wondered if Dawkins's simulated evolution could produce for me a biomorph that resembled a Christmas tree. In 10 generations I had this:

(one of the generations included a mutation that introduced bilateral symmetry).

In 20 generations I had this:

Of course, this is no more than an illustration of the power of the combination of random mutation with non-random selection. The mutation process explores the branching possibilities; the selection process prunes those branches. When I play with the Blind Watchmaker program, it's me who makes the selections: which one looks most like a Christmas tree? In a real evolutionary process, the selection force is the likelihood of the organism to survive and reproduce. For one of the Dawkins's biomorphs to survive and reproduce on my computer screen, success means meeting with my approval. In nature things are a lot more complex, but just as brutal. Organisms that fail to survive long enough to reproduce reach the end of the line.

Even though the selection stage is the final arbiter of which organisms survive and which do not, the random variation stage is what creates the novel choices from which the selection is to be made. Without variation, without chance, there would be no variety in life. Indeed, it is overwhelmingly likely that there would be no life at all.

Random logic

The essential processes of biological evolution are random variation, inheritance from one generation to another and selection. There's a case to be made that processes akin to these are essential to all creative activity.

In contemplating the framing of my next paragraph (yes, this is a self-referential piece of writing) I am bringing together in my head multiple possible ways of expressing the thought I wish to express. Like Dawkins's biomorphs, I have in my mind a few different versions of the thought itself, and I am trying to choose the one that fits best. I consider various ways of expressing that thought, sometimes combining the best aspects of two phrases, and finally I select the wording that pleases me most. Later, on revision, I will see if the paragraph is still fit for purpose and I will allow it to survive or I will sacrifice it.

Similar evolutionary processes can work for different kinds of creative activities. It might take some ingenuity, but most designs can be expressed in some form of digital code to create a domain-specific 'genome'. If the design can be parameterized, and reduced to a digital form, then a computer can be programmed to create mutations of the design. If that can then be coupled to an effective selection process, we can allow a genetic algorithm to evolve, generation by generation, better versions of the design.

NASA has developed an encoding for airplane shapes, called OpenVSP. In a 2014 paper, researchers from University College, Dublin, described a project to 'evolve' an aircraft.[134] This project coupled NASA's encoding for airplane shape with simulations of airflow to produce optimized wing shapes. The parameters in NASA's encoding were varied, and the best solutions were selected for maximum lift and minimum drag.

Earlier than that, for the 2006 space mission Space Technology 5, NASA had used evolutionary design to create antenna designs for the three satellites involved in the mission. The requirements were unusual, and the resulting antennae looked unlike anything that human engineers might have considered: the computer algorithm had explored a wider solution space, and had arrived at the solution more quickly. Then, when the requirements changed, the computer design approach was able to adapt the design very quickly. Increasingly, such genetic algorithms are becoming viable ways of approaching real design tasks, and it seems likely that evolutionary approaches will have an increasingly important role to play in mainstream engineering.

It's not just physical structures that can be evolved into better forms by following a genetics-inspired approach. Part of the surge of effective artificial intelligence (AI) applications in modern technology results from approaches that allow computers to find their own algorithms, often using mechanisms inspired by biological evolution. Machine learning is a complex business, but

---

[134] *Evolving an Aircraft Using a Parametric Design System*, J. Byrne, P. Cardiff, A. Brabazon, M. O'Neill, 2014.

randomness plays an essential role in allowing a computer to learn to carry out meaningful problem-solving tasks.

An important class of machine-learning systems involve neural nets, software so named for their conceptual similarity to the neural structures of the brain.[135] These are highly abstract networks of relatively simple computational nodes. When large numbers of these nodes are properly linked together, the resulting network can accept inputs and generate outputs in a flexible way that can emulate complex logic. Unlike conventional computer programs, though, the logic that links input and output is not explicitly specified by a coder, but can evolve through a simulated learning process. The connection between input and output is a function of numerical weights attached to the nodes and the links between them. In this way, the neural net mimics the linkages between neurons in the brain.

At the outset, a neural net with 'untrained' weightings will produce meaningless outputs, but through a process involving randomization and a reward mechanism, any settings that might by chance generate promising outcomes can be rewarded by adjustments to the weightings given to the nodes and links. The effect of these rewards for those promising settings is to be allowed to propagate—those successful settings are more likely to be preserved in successive generations of the algorithm. Through many iterations of training runs, the weights of the elements of the neural net are ever more finely adjusted, and they start to produce outputs that begin to resemble what's required, even if in the early stages of the process the results seem nonsensical.

Janelle Shane, an AI researcher and writer, keeps a playful blog called *AI Weirdness*[136] where she documents her experiments with AI technology. In one of these experiments, she supplied a neural net with a training data set consisting of a list of 7,700 paint colours and their marketing names from the paint manufacturer Sherwin-Williams. Some examples of the inputs: Decorous Amber, Majolica Green, Pewter Tankard, Downing Straw. Here are some samples of what the AI algorithm came up with, as it went through generation after generation of refinement. Bear in mind that the starting point is a complete

---

[135] Although this is a fair comparison, our artificial neural nets are vastly smaller than human brains: they are more on the level of the brain of an earthworm.

[136] AIWeirdness.com is highly recommended for both information and entertainment. In one of her experiments, Janelle Shane gave a neural net a training data set consisting of a list of pick-up lines. She chose one of the computer-generated outputs as the title for her book: *You Look Like a Thing and I Love You*.

blank canvas: even forming strings of characters that look like words is an achievement.

- Early outputs were boringly meaningless, and barely pronounceable:
  - Caae Brae, Caae Blae, Saae Ble, Reree Gray, Conk Green
- With a tweak to increase the 'creativity', things got a little weirder:
  - Bylfgoam Glosd, Gorlpateehecd, Iroeee CerMowt, Donarf
- But examples from later in the process suggest the AI was getting the idea, but clearly lacked the cultural filters that a human would apply when selecting names for paints:
  - Navel Tan, Snader Brown, Burf Pink, Rose Hork, Grass Bat, Turdly[137]

In Shane's description of the process it's clear that even though the neural network is doing the job of producing the colours and their names, the learning process requires human oversight. To get anything like useful output requires some human intervention, both in the selection of the training data, and in tuning the overall parameters. However inept the final results might seem, we should not lose sight of the fact that even this is a considerable achievement. These amusing colour names are the result of a process that starts with no information at all about colours, and no knowledge of word construction other than that one letter follows another. All its flounderings are directed only by a scoring mechanism that judges the ever-improving outputs.

The AI's algorithm feeds back on itself, generation by generation, and tunes itself to produce better and better outputs, as measured by a scoring rule. It has no conceptual understanding of *why* it works that way. The neural nets contain all the logic to produce plausible outputs, but in this case, as in the case of most neural nets, there is no way for us to read off an easily understandable explanation of the principles being used in coming up with the names. The nodes in the neural net are connected in complex ways and we can verify that they are functioning correctly, but there is no representation that we can see in the system of the *concepts* of colour or names, or even those of word formation. We cannot ask the AI why it makes the choices it makes.

Shane tells the story of an AI image recognition algorithm that started seeing sheep in images where there were clearly no sheep present. These were all

---

[137] Her neural net was also producing the matching colours, not reproduced here.

images of 'lush green landscapes'. The AI had been trained on a set of images where sheep were always pictured in such verdant surroundings and it had associated the label 'sheep' with the landscape and not the animal. Show it a sheep in an unconventional location, a car or a living room, and the AI was likely to declare it to be a dog or a cat. Janelle Shane's neural nets are structures that emerge through the interaction of random variation and harsh selection, and as they iterate towards more and more credible solutions, they create internal logic that is opaque to outside observers. That's not unlike the process of natural evolution, where organisms are subjected, over many generations, to forces of evolutionary pressure and respond with fascinating and often complex solutions, but leave us without obvious explanations of exactly how those solutions came to be.

The natural world is full of wonders of adaptation. Orchids mimic bees; archerfish spit at and so capture insects flying above the water; spiders weave webs with geometric precision. Trying to ask why these solutions evolved is like asking *why* about the outputs of Janelle Shane's experiments. We can see that there is logic there, but it requires deep effort to uncover that logic. The world is a rich mixture of chaos and pattern, and we are inquisitive creatures. We crave explanations and meaning. We notice the patterns lurking in the variety of the world and we ask the question *why*.

Answer to:
Noisy colours

Most of these different colours of noise are so-called because of an analogy with visual light, but not all of them.

- Pink noise is more weighted to low frequencies than white noise. It's also known as $1/f$ noise, as the power at any frequency is in inverse proportion to the frequency.
- Blue noise is more weighted to high frequencies than white noise. The power at any frequency is in direct proportion to the frequency.
- Green noise is an informal term used to refer to noise with more power in the middle frequency range.
- Brown noise, technically, corresponds to $1/f^2$ noise and is also sometimes referred to as red noise. The name used is derived from Brownian

motion, the chaotic motion of small particles on the surface of a liquid, rather than the colour (brown is not a colour which appears on the visual spectrum).

• Since black, visually, is the absence of light, black noise is a jokey way of referring to silence.

The fifth duality: Disruption–Opportunity

In 1922 Ernest Hemingway, aged 23, was a barely published writer and reporter for the *Toronto Star*, in Lausanne to cover a peace conference. Since he was missing his wife, Hadley Richardson, who was in Paris, they arranged that she would join him. She packed a suitcase with his writing: notes, originals, and carbon copies for a novel in progress. Then, disaster! In the Gare de Lyon, the suitcase went missing. Everything was lost. Some say that Hemingway decided after this loss to write more economically, which became his characteristic style. Mistakes, accidents, and chance events disrupt our flow. But sometimes the disruption is for the good. A shake-up can create new opportunities.

Most small businesses fail: one estimate is that 60 per cent of start-ups in the United Kingdom fail within five years. Are those wannabe start-up heroes poor deluded fools, or brave entrepreneurs? Every firm was once a risk that some brave company of fools undertook. The commercial world depends on those risk-takers.

When the road ahead is blocked, we notice side paths that we might otherwise pass by. Disruption forces us to explore. It's unsettling to reflect on how many technical advances result from wars and other crises, and even in peacetime, funding for research seems to come a little more easily when there seems to be a possible military application. The internet itself was a creation of the US Defense Advanced Research Projects Agency. The resilience of the internet was funda-mental to its initial design: the US military needed a network that could survive nuclear attack, and designed one that could survive the loss of any part of it.

Art is a collaboration between the artist and the audience. The appreciation of art occurs when elements in the artwork collide and interact with the ideas, memories, and associations that are already, and uniquely, present in the mind of the recipient. In that moment of conjunction, something new is made: an experience that is particular to that artwork, that person, that time and place. It's the product of chance.

So this is the fifth duality. Chance is scary, but chance is also thrilling. Fortuitous conjunctions and random combinations are at the heart of creative processes whether in art, commerce, science, or nature. Chance is a necessary and welcome part of life.

PART 4

Taking Charge of Chance

Florence Nightingale: The lady with the stats

The main end of statistics should not be to inform the government as to how many have died, but to enable immediate steps to be taken to prevent the extension of disease and mortality.

Florence Nightingale

Florence Nightingale (1820–1910) is hardly an unknown figure. As a child I knew her story—or thought I did: the pioneer of nursing, sent to administer to those wounded in the Crimean War (1853–1856), where she became famous as an almost saintly dispenser of care and comfort. She was the Lady with the Lamp whose presence alone, as she moved through the night-time wards of the hospital at Scutari in Constantinople, brought ease to the suffering soldiers. In the public eye she became the very ideal of a nurse. Her name is synonymous with nursing. The buildings adapted by the UK Government into makeshift hospitals at the height of the Covid-19 pandemic were dubbed 'Nightingales'. To this day, in the United States, nurses recite the Nightingale Pledge at their pinning ceremonies on entering the profession.[138]

Nightingale had always been well connected politically—her posting to Scutari had been arranged by Sidney Herbert, a family friend and Secretary of State for War. But in Britain, starting during the war and continuing after her return, she became a celebrity. The public image created around her ensured that her voice could never be ignored. The public knew the perfect nurse—the one in the books I read as a child—while behind the scenes she was a confidant of politicians and administrators, campaigning for public health.

I later learned that there was much more to her story than nursing: she was also a statistician. After the war in Crimea she studied the numerical records of deaths in that conflict, and from her analysis created persuasive diagrams to communicate clearly to politicians and others that the catastrophic death rates in the Crimean War were caused in very great part not by the conflict itself, but by disease, especially among those wounded and sent to hospital. Her diagrams,

---

[138] In its original form, those taking the Nightingale Pledge declare: 'I solemnly pledge myself before God and in the presence of this assembly to pass my life in purity and to practise my profession faithfully. I shall abstain from whatever is deleterious and mischievous, and shall not take or knowingly administer any harmful drug. I shall do all in my power to maintain and elevate the standard of my profession and will hold in confidence all personal matters committed to my keeping and all family affairs coming to my knowledge in the practice of my calling. I shall be loyal to my work and devoted towards the welfare of those committed to my care.'

early infographics, make this distinction plain. They also show the rise and fall in the deaths from disease through the war, starting in April 1854 (the war had started in October 1853), rising to a maximum in January 1855 and falling away to March 1856.

While Nightingale's diagrams illustrate effectively the severity of the death rate arising from disease, they do not in themselves explain why there was so much disease; nor do they account for the improvements in death rate after January 1855. Nightingale had thought initially that the soldiers were being sent to hospital too late—that they were already dying when they arrived there—and she also laid some of the blame on poor nutrition. The way in which diseases spread was not at that time well understood: it was only in the 1860s that Louis Pasteur started doing formal experiments to investigate the links between germs and disease.

After the war, when Nightingale studied the statistics with the support of William Farr, she became convinced that the primary factors behind the appalling death rate were inadequate hygiene and poor sanitation.[139] The hospital at Scutari had been, in her later words, 'a pesthouse'. Sanitary improvements had been undertaken during the course of the war by John Sutherland as head of a Sanitary Commission sent to Scutari in 1855, and these had helped conditions there. Nightingale saw that she had missed many opportunities to improve matters herself.

This realization drove Nightingale's passion for the cause that followed: the improvement of public health through better sanitation. She studied the death rates in military barracks in Britain, comparing them to the rates in equivalent civilian populations, and found that the soldiers experienced a death rate twice as high, something she now ascribed to a lack of cleanliness and otherwise poor conditions. She retreated from the world physically, but she was by no means isolated. Through correspondence and other writings she continued to exert influence on public health matters. She contributed substantially to the design of St Thomas's Hospital in London, built as seven 'pavilions' with connecting walkways, large windows, and separation of clean and dirty areas, a design intended to address the issues that she now knew had plagued Scutari. She wrote for the public, too: her book *Notes on Nursing: What It Is and What It Is*

---

[139] At that time, the very words hygiene and sanitation referred to health only in broad terms, and were not specifically associated with cleanliness. Hygiene derives from the Greek *hygies*, 'healthy, sound, hearty', and sanitation from the Latin *sanitas*, 'health'.

Not became very popular; it is more about the environmental conditions appropriate for the care of patients than it is about specifically medical matters.

In the chaos of war it had not been possible to discern, with any clarity, the factors that had caused the death of so many soldiers in the hospital at Scutari. But adequate statistical records had been kept, and it was the study of these records that persuaded Nightingale of the central role played by hygiene and sanitation. Statistics bought clarity from chaos. Understanding how diseases spread became fundamental to creating healthy living and working conditions.

Time and again, we see that the first effective step in combatting illness, which at first sight appears to strike randomly, is to analyse the numbers and create a model through which the underlying mechanisms can be investigated. This brings the understanding that then shows the way to effective approaches to prevention and treatment.

Taking charge of chance

There is no such thing as absolute truth and absolute falsehood. The scientific mind should never recognise the perfect truth or the perfect falsehood of any supposed theory or observation. It should carefully weigh the chances of truth and error and grade each in its proper position along the line joining absolute truth and absolute error.

Henry Augustus Rowland

Fighting back

It's one thing to have the facts, to know the odds, to understand the risks; it's quite another to be able to exercise some control over chance. It often feels that we are passive and powerless victims of fate. On a personal level, accidents frequently disrupt our plans and force us to accommodate random mishaps. At a larger scale, natural disasters cause immense loss and apparently meaningless havoc. The political storms that whirl around us seem scarcely more understandable.

But we are not passive creatures, doomed simply to accept the luck we are given. By our actions we can not only mitigate the effects of bad luck, but also arrange our lives and our affairs so as to put ourselves in a position to exploit what good luck might come along.

Randomness is not all of one kind. By making observations and taking measurements, we can understand the patterns of randomness. From this tentative understanding we can guess at underlying truths, and build plausible models consistent with the patterns of our observations. This job of working backwards, of discerning possible explanations for what we see, and then testing and validating the models, is a job for statisticians.

The focus of this book is chance, and the mathematics of chance is probability. Probability calculations start with a model and work forwards. We model a coin flip as having two equally possible outcomes, and take that as a starting point. Statistics reverses that logic. If we see a coin landing heads 13 times out of 20, what can we say about the underlying model? Is the coin bent or true? Statistics starts with a jumble of observations, and works back to discover a plausible model of the underlying truth. In a world of randomness, statistics offers a path to discovering, sometimes only tentatively, models of explanation.

The models may be no more than provisional guesses at the truth, but we can nonetheless use them to make effective change in the world. Florence Nightingale had to reject her early theories of why so many soldiers died at Scutari, but once she had concluded that the cause was poor sanitation, she was able to translate that understanding into effective programmes of action.

Statistics, as a discipline, is an even more recent arrival than the theory of probability. Francis Galton, the inventor of the quincunx, that pinball toy he used to demonstrate that random processes can give rise to stable patterns, is one of the pioneers of mathematical statistics. Galton was a man of extraordinarily diverse accomplishments. Born in 1822, the grandson of Erasmus Darwin (and so a cousin of Charles Darwin), his interests spanned topics from meteorology and geography to heredity, fingerprints, and hearing tests. And he stands at the head of a line of academic descent of eminent and hugely influential statisticians. Karl Pearson and Ronald Fisher built on his work: together they are responsible for much of the edifice that is classical mathematical statistics, encompassing regression analysis, hypothesis testing, and many other techniques upon which statisticians rely.

Our respect for Galton's unquestionably important work must be balanced by a recognition of his racism and his role as the founder of the pseudoscience of eugenics. Galton had been struck by the force of the ideas in Charles Darwin's book *The Origin of Species*. From that idea of natural selection he derived and promoted the idea that human populations could, and should, be genetically 'improved' by programmes of what would amount to selective breeding. Some of his academic descendants, Pearson and Fisher among them, followed the same line of thinking. These figures challenge us: respect for their undoubted intellectual brilliance and achievement sits alongside abhorrence of their racist ideas.[140]

But there is a deeper and more noble impulse behind statistical analysis: to understand the truths that lie beneath the appearance of randomness, and so to equip and empower us all to make better choices for better lives in a better world. When epidemiologists study an outbreak of disease, the early statistics collected about the transmission and progression of the disease open the gateway to understanding. That understanding shows us options for intervention that might be effective in containment, treatment, and prevention.

Just as medical advances have dramatically reduced the impact of disease on the world's population, so too have improvements in health and safety reduced the risk of accidents. Physical improvements to cars and other machines, attention to food and drug safety, and process improvements to systems of all kinds have helped us all to live more safely. And if technological innovation comes with some risk of its own, we should not discount the extent to which technological solutions can mitigate that risk, and it is often through statistical analysis that the routes to the best solutions are found.

Not all risks can be prevented, though. Despite our best efforts, we will never be able to eliminate all the blows of chance. Any one of us, at any time, can be struck by catastrophic misfortune. Businesses fail, mistakes are made, buildings catch fire, accidents happen. But the law of large numbers can help. If we can agree to pool our individual risks, we can make everyone safer. This is what an organized society does. We pay our taxes, and in return we expect to receive the shelter of the State when, through no fault of our own, we are the victims of random misfortune. To go further, we take out insurance contracts to provide cover for ourselves and the things we value. At a predictable cost, paid regularly, we gain the peace of mind that, when and if catastrophe strikes, we will be helped. In this way, we turn individual vulnerability into collective security.

---

[140] On 9 June 2020, Rothamsted Research announced that the name of one of their buildings, 'Fisher Court', was to be renamed 'AnoVa Court'. AnoVa (analysis of variance) is a statistical technique developed by Fisher's team at Rothamsted Experimental Station, where he worked for 14 years. The name of the building is now a recognition of the work, not the man.

CHAPTER 14

The Weight of Evidence

In our reasonings concerning matter of fact, there are all imaginable degrees of assurance, from the highest certainty to the lowest species of moral evidence. A wise man, therefore, proportions his belief to the evidence.

David Hume, *An Enquiry Concerning Human Understanding*

Foul or fair?

You visit your good friend Pat. You are old board-game-playing buddies, and you're intrigued to see that a dice has been left out on the coffee table. Pat heads off to make the coffee, and you idly play with the dice. You notice something curious, and so you roll a few more times, this time noting the numbers that appear. This is what you've written down when Pat arrives back with the coffee:

6 3 6 1 6 5 5 6 1 3 5 3 6 6 5 6 6 6 6 4

'Ah, I see you found my crooked little friend. I suppose you've guessed that it is rather fond of rolling sixes. In fact, that dice has been weighted to roll a 6 half of the time.'

'I guessed something like that.'

'But I have a question for you.'

'Yes?'

'Tell me, how soon did you suspect something was up? How many rolls did it take before you started to think that something was wrong with that dice?'

Think about it. Is it very odd that two 6s should appear in the first three rolls? Surely not, that must happen occasionally with a perfectly fair dice. Three 6s out of five rolls? Four out of ten? Ten out of twenty? A run of four 6s? How much evidence do you need before you start to smell a rat? How much more evidence must you see before you are certain enough to call foul?

Discussion in the course of this chapter.

Proof and probability

The word probability, so central to this book, deserves some attention. It comes from:

probable (adj.)

> 'likely to happen or be the case' from
>
> **probable** (Old French) 'provable, demonstrable'. From
>
>> **probabilis** (Latin) 'worthy of approval, pleasing, agreeable,
>>
>>> acceptable; provable, that may be assumed to be believed, credible'. From
>>>
>>>> **probare** 'to try, to test'.[141]

Going by the etymology, then, probability is a measure of how 'provable' something is. And that leads us to an area where proof is a central concern: the law.

Crime and probability

When uncertainty simply means a lack of knowledge, the past can be as uncertain as the future. In the board game *Cluedo*,[142] the players compete as detectives to solve a murder: three cards are chosen at random, one from each of three small packs representing, respectively, the murderer (one of six possibilities), the weapon used (one of six), and the location, one of the nine rooms depicted on the game board, which represents the ground floor plan of a mansion. These cards, faces hidden from all players, are slipped into an envelope and set aside. The aim of the game is to deduce which three cards have been removed from circulation and thus to 'solve' the murder. Any player who has pursued the logic to the point of certainty can make an accusation: 'I accuse Colonel Mustard, with the dagger, in the ballroom'. If, on checking the envelope, the accusation

---

[141] All etymologies in the book are from etymonline.com.
[142] The game is known as *Clue* in some countries.

proves correct, that player has won. If the accuser is mistaken, the rest of the players continue until a correct accusation is made.

The game relies mostly on logic and elimination of possibilities, and only a little on reasoning about probability, but there is one aspect of interest to us. This is that the game play concerns learning about a random event that has already happened. The randomization of the cards happens at the outset, and the players start from a position of ignorance, where all 324 combinations of murderer, weapon, and location are equally possible. As each player acquires evidence, they eliminate more and more of the combinations, until only one remains: the solution to the murder. But the probabilistic reasoning the players use is not about the future, but about the past, as they deduce what must have happened when the cards were randomly selected at the start.

Just as *Cluedo* players start from a position of maximum uncertainty and use clues as evidence to guide them to certainty, so it is in the law courts. The judge or jury must proceed from an unprejudiced starting point, weighing the evidence presented and arriving at a justified determination of the truth of the matter. The courts and the jurists who sit in them are there to ensure the resolution of uncertain matters and provide settled judgements. Judges and juries, presented with incomplete and sometimes biased evidence ('partial' in both senses of the word) from potentially unreliable witnesses, must come to a version of the truth, and must do this in a way that satisfies society's expectation of proof. They have to answer the question 'What are the chances of that?' where 'that' refers to the charges levelled at the accused person.

A formalized, if not quantified, approach is needed. How 'provable' is the case against the prisoner in the dock? How probable? Since ancient times, courts have been convened to decide such matters, and have relied on formal protocols for weighing the evidence before them. Some of these protocols foreshadow modern ways of thinking.

For example, the record of a tribunal in ancient Egypt from around 2200 BCE regarding a will dispute notes that evidence was required from three honourable and credible witnesses, each of whom had to take an oath before testifying. If we treat those witnesses as independent, then in a crude use of probability theory we can say that, if there were a 20 per cent chance that a witness is false, then there would be only a 0.8 per cent chance[143] that all three independent witnesses would give false testimony. The more witnesses there are in agreement, the greater the confidence in the truthfulness of them all.

---

[143] For independent events, we multiply probabilities: $0.2 \times 0.2 \times 0.2 = 0.008$.

The Talmud, the ancient source of Jewish law, distinguishes between the levels of evidence required for different kinds of trial. In a capital case, circumstantial evidence was never enough: 'Perhaps you saw him running after his fellow into a ruin, you pursued him, and found him sword in hand with blood dripping from it, whilst the murdered man was writhing. If this is what you saw, you saw nothing.' If a step of logical inference was needed to link the evidence to the crime, it counted for nothing. By contrast, in civil cases that could be remedied with monetary compensation, a lower standard of evidence was demanded. One or sometimes two witnesses would be sufficient. Though no explicit measurement of probability was involved, the different levels of proof that were required were in effect measures of different levels of certainty/uncertainty.

The Talmud also recognizes the usefulness of presumption—that which can be expected in the absence of evidence to the contrary. A person is presumed to be alive in the absence of evidence of their death. The term of a pregnancy is presumed to be nine months. 'What normally happens' is given due weight. As we will see later in this chapter, the explicit incorporation of such prior knowledge is an essential part of how we now incorporate evidence in reasoning about probability.

Burden of proof

Different levels of proof are required for different kinds of trial. In English civil (as opposed to criminal) law, the standard is 'balance of probabilities'. Simply weigh one side against the other and decide where the evidence falls most heavily. For criminal conviction, as in the Talmud, a much higher standard of proof is necessary. The case must be proved 'beyond reasonable doubt'. The jury members must be 'sure'—whatever that may mean.

Cornell Law School's Legal Information Institute distinguishes these different standards of proof used in the United States:

- **Some (credible) evidence**: Used, for example, in administering prison discipline
- **Substantial evidence**: Used by appellate courts in assessing findings of an inferior court
- **Reasonable suspicion/indications**: Used to justify police 'stop and search' actions
- **Reasonable belief**: Used to justify vehicle searches after arrest of a suspect

- **Probable cause**: Used to justify police actions such as arrest or search of premises
- **Preponderance of evidence**: More likely than not; the burden of proof in a civil trial
- **Clear and convincing evidence**: Used in cases relating to financial crimes such as fraud
- **Beyond reasonable doubt**: Used in criminal trials.

While these standards form an approximate scale of increasing certainty, no numbers are attached to these levels, other than for *Preponderance of evidence*, which is usually taken to mean more than 50 per cent probability.

The law distinguishes between direct evidence and circumstantial evidence. If you need to take a step of logic, an inference, to connect the evidence to the crime, and a different interpretation is possible, even if very unlikely, then you are dealing with circumstantial evidence. And it is in the assessment of the likelihood of alternative explanations for circumstantial evidence that probability plays a role.

Law courts have been reluctant to embrace systems of evidence based upon mathematics and explicit quantification of probabilities. But courtrooms are not the only arenas in which people are judged, where evidence must be weighed and decisions must be made based on incomplete or unreliable information. Artificial intelligence algorithms play an increasingly important role in such matters as credit rating and recruitment, which traditionally rely on human judgement. The algorithms are based on mathematical approaches to weighing the evidence provided as data inputs. As in the law courts, we need to consider carefully how much trust to place in these 'witnesses', and how to combine their testimonies to lead us to a satisfactory judgement.

Inverse probability

In his *The Art of Conjecture* published in 1713, Jacob Bernoulli begins to consider the question at the heart of statistics, which reverses the logic of the probability questions that others had been debating. When we calculate the probabilities associated with rolling dice, we think in a forward direction. The theory of dice-rolling and of calculating the probabilities of complex roll combinations is based on an underlying model that makes assumptions, for example that each face of the dice appears on average equally often. This approach works well when there is plausible independent justification for such assumptions.

But what if we don't know the model, only the observed consequences? Can we figure out the model by working backwards? Bernoulli imagined an experiment involving balls in an urn. (Statisticians love that sort of thing.) If we have an urn containing five balls, black and white, but we don't know how many there are of each colour, we can try some experiments to gather some evidence. If we blindly take a single ball out, note its colour and return it to the urn, what can we say about how many black and how many white balls are in the urn? With only one item of evidence, we can't say much—although it would be reasonable to suspect that whatever colour ball we drew, that colour might be more likely to turn out to be in the majority. But suppose we did this 50 times, each time replacing the ball, and on looking at our notes we found there had been 8 times when a white ball had been taken from the urn, and 42 times when the random ball had been black. What could we say then? There'd be good reason to guess that the mix of balls in the jar was four black to one white. The evidence of 50 repeated drawings from the urn does not provide absolute certainty, but it is strongly suggestive. I've used some vague words there: 'good reason', 'strongly suggestive'. Can we make that vagueness more precise? Can we measure it or put a number to it? We can. We can reverse the direction of thinking about probabilities and calculate the chances of each possible combination of five white and black balls in the urn in the light of the evidence we have. Statistics can help us build, from our observations, a plausible model of what lies behind those observations, and to say how confident we are in our conclusions.

Sometimes the shape of the model is clear, and only a little detail is needed: how many black balls are there, how many white balls? But most often there is very little clarity to start with, even in knowing what factors should be considered. The number of cases of lung cancer increased dramatically in the early twentieth century. What was the cause? Tarmacking of roads? Factory working conditions? Cigarette smoking? Reasoning backwards from observations is not always straightforward. This working from evidence to an understanding of the probabilities involved was once called the problem of 'inverse probability'. Nowadays, though, we know it as Bayesian probability, and we owe that name to an English clergyman.

Thomas Bayes

The Reverend Thomas Bayes was born around the year 1700. He studied logic and theology at the University of Edinburgh and served as a Presbyterian minister in Tunbridge Wells from 1734 until 1752. In 1763, two years after his death, a paper he had written, 'An Essay towards solving a Problem in the Doctrine of Chances', was published. This paper addresses a specific question in probability, but before Bayes tackles the main problem, he establishes some preliminary results, including the result that has since become famous as Bayes's Theorem. In short, this theorem provides a way of combining a 'prior' degree of belief in some proposition with a new element of evidence that may alter our belief. The result is a 'posterior' degree of belief that incorporates the new evidence. Here is the key vocabulary if you want to talk the Bayesian talk, describing the process of incorporating a new piece of evidence into our understanding.

- **Prior probability (or often just 'priors')**: An assessment of the probability of the event we're interested in, before we take the new evidence into consideration
- **Evidence**: New information that shifts our belief about the probability of the event
- **Posterior probability**: The assessment of the probability of the event after incorporating the new evidence.

Getting technical: Bayes's Theorem

The formula for Bayes's Theorem:

$$P(A\,/\,B) = \frac{P(A)P(B\,|\,A)}{P(B)}$$

The notation $P(A|B)$ is called a conditional probability and should be read as:

'the probability of A, given B'.

So to interpret Bayes's Theorem, we should read it as:

The probability of A, the event we're interested in,
given B, the evidence,

and it can be calculated as

the (prior) probability of A without any further evidence,

multiplied by

the relative likelihood that the evidence B would be present if A were true

$$\left(\frac{P(B\,|\,A)}{P(B)}\right).$$

continued

Getting technical: Bayes's Theorem *Continued*

Let's take an example:

We can use L as a shorthand for the condition that an individual in a certain population has lung cancer

and

S can mean that the individual is a cigarette smoker.

Then suppose we know that 90% of lung cancer patients are cigarette smokers, so the probability of a lung cancer patient being a smoker is 90%; we can write:

$P(S|L) = 0.9$.

If, among the population being studied, 5% are lung cancer patients:

$P(L) = 0.05$.

And if 20% of that same population are smokers,

$P(S) = 0.2$.

Then $P(L|S)$ is the chance that a patient from that population has lung cancer, if you know that they smoke cigarettes.

From the formula we get:

$$P(L|S) = \frac{0.05 \times 0.9}{0.2} = 0.25 = 22.5\%$$

In other words, for this population, if you know that a patient smokes cigarettes, then the chance that they will be found to have lung cancer increases from 5% to 22.5%.[144]

The loaded dice

Do you remember Pat's loaded dice that featured in the brainteaser at the start of this chapter? We can use Reverend Bayes's formula to weigh up the evidence of those dice rolls.

Perhaps Pat offered you the choice of two identical-looking dice: one of them is the loaded dice which rolls a 6 half the time, and the other is perfectly fair and rolls a 6, as you'd expect, one in six times. Pat swaps the dice from hand to hand, and you randomly choose one of them. So the prior probability that you get the fair dice is 50 per cent. How much can you learn about which of the two you have ended up with?

---

[144] Note: this logic does not in itself prove that smoking cigarettes causes cancer. A later section (starting on p.273) explores the story of how the links between smoking and lung cancer were established.

Roll the dice once: it's a 4. All we care about is whether it is a 6 or not, so that counts as a non-6. Does that tell us anything at all? You might say no, that you need more rolls to start jumping to conclusions. But suppose that this is the only evidence you could get. Imagine that Pat is a comic book supervillain and your fate depends on choosing the fair dice. You get one roll only and then you must choose; so you'll take any clue you can. Your one roll came up 4: does that not suggest, even slightly, that the dice you're holding is more likely to be the fair dice than the crooked one?

This is where Bayes's wonderful formula comes to the rescue. Work through the numbers, and you'll find that evidence of the single roll has increased the chance that you have picked the fair dice from 50 per cent to 62.5 per cent. In the jargon of the Bayesian statistician, your *prior probability* of 0.5 has been superseded by a *posterior probability* of 0.625. Stick with that dice.

Getting technical: putting Bayes to work

To calculate the effect of a single roll of Pat's dice, we can use Bayes's formula:

$$P(A/B) = \frac{P(A)P(B|A)}{P(B)}$$

In this case:

 A means 'the dice is fair'

 B means 'a non-6 has been rolled', whether from a fair or a crooked dice.

Let's work through the numbers needed:

 $P(B|A)$ is the probability of rolling a non-6, for a fair dice. This is $5/6$.

 $P(A)$ is the *prior* probability that the dice is fair. We made a random choice between two dice; so this is $1/2$.

 $P(B)$, the overall probability of rolling a non-6, whichever dice you chose, is a bit more complicated to calculate, as we need to cover the cases of both a fair and a crooked dice.

 To get the probability of the dice being fair and a non-6 being rolled, we must combine the probabilities of the dice being fair ($1/2$) and the roll not being a 6 ($5/6$). Multiplying these together comes to $5/12$.

 To get the probability of the dice being crooked and a non-6 being rolled, the crooked dice must have been chosen ($1/2$ chance) and the roll must be something other than a 6 ($1/2$ chance). Multiplying these gives a $1/4$ chance.

 Add together the chances of the two ways of getting a non-six, and we have $8/12$.

 Put all these numbers into Bayes's formula to get $5/8$.

 In other words, if something other than a 6 appears, we have calculated that there is a $5/8$ or 62.5% chance that the dice we have chosen is the fair one. Using the evidence of one roll, we have refined the probability from 50% to 62.5%.

258 | WHAT ARE THE CHANCES OF THAT?

Bayes's formula is often boiled down to a number called the likelihood ratio: it's the ratio of the probability of one outcome to that of another. In this example, the likelihood that you chose the fair dice beats the chance that it was the loaded one by a ratio of 5 to 3. If you multiply the prior odds (in this case 1 to 1) by the likelihood ratio, you get the posterior odds (5 to 3 on[145]).

If Pat allows you to keep rolling, you can get more evidence. Eventually you will arrive at a point where you can have enough evidence to decide that dice is almost certainly fair (probability of being fair is close to one) or crooked (probability of being fair is close to zero). How confident must you be to treat it as the truth? That's a matter of judgement, but regardless of how stringent a criterion you set, the more rolls you make, the closer you can get to that criterion, and satisfy any standard short of absolute certainty.

Suppose you roll ten times, getting this sequence of 6s and non-6s.

3 1 **6** **6** 4 **6** 3 3 1 **6**

What are the chances that that particular sequence would be rolled by a fair dice? Sparing you the details, it's about 1 in 60 million. What are the chances that the crooked dice would roll that sequence? About 1 in 16 million.[146] What does Bayes's formula say about that? That there is only around a 21 per cent chance that your dice is the fair one. If you see that sequence, crookedness is around four times more likely than fairness.

Gather more evidence: roll the dice ten more times, extending the sequence to this 20-roll run:

3 1 **6** **6** 4 **6** 3 3 1 **6** 2 **6** 5 **6** 1 **6** 4 **6** **6** 3

What is the chance of rolling that specific sequence with the fair dice? It is about 3,656 trillion to 1—an extremely small number. And what's the chance of rolling that particular sequence with the crooked dice? It's 51 trillion to 1, which is still a ridiculously small number. Plugging these into Bayes's formula, we can calculate the chance that the dice is the fair one: around 1.4 per cent. The likelihood ratio in favour of crookedness is around 70.

---

[145] Recall that odds-on are the odds in favour of an event, and are mostly used where the probability is greater than 1.

[146] You might argue that not the precise sequence, but only whether each roll was a 6 or not is relevant here. You might go further and argue that the order in which 6s and non-6s appeared was irrelevant. You'd have a fair point. Whichever way you reason it, working through Bayes's formula in all three cases comes to the same chance of your having chosen a fair dice.

In passing, an observation: both of those exact sequences were highly improbable. You'd expect that to be the case when looking at the chances of a particular sequence of random events. But because we are comparing the *relative* probabilities of two very improbable sequences, the comparison is a sensible one. Thinking only about the loaded dice, you might think that since the chance of getting that run of numbers is so small (51 trillion to one against), you can dismiss it as a possibility. But you can't do that, because the chances that that run of numbers came from a fair dice are even smaller, and one or the other of those events has in fact taken place. Highly improbable events occur all the time, and to assess them, we must always compare them to the alternatives, which may also be highly improbable. We will have reason to revisit this logic soon when we return to the courtroom.

Context makes a difference

A Bayesian approach takes explicit account of the prior probability, in our case the chance that the dice is fair or loaded even before we have any evidence. Pat offered you a 50/50 chance of choosing one dice or the other, but if, say, you had been offered the choice of dice from a box of ten, of which nine were loaded and only one was fair, the analysis and the calculations would have gone differently. In that case the prior probability of a fair dice would have been only 10 per cent, and the posterior probability after ten rolls would have been around 3 per cent (and after 20 rolls around ¼ per cent). It's not always this easy to know what your prior should be. If you're in an informal game of craps, what's the prior probability of a dice being fair? It has to be a subjective judgement.

Meanwhile, back in the courtroom

When assessing circumstantial evidence, a court is essentially weighing probabilities: testing the inferential logic that connects the evidence to the crime. But even though Bayesian logic provides a coherent and mathematically sound way of combining probabilities relating to evidence, this has not been fully embraced by the law courts. In the United Kingdom, though, the courts will recognize the concept of the likelihood ratio, and interpret this verbally as follows:

Likelihood ratio	Verbal interpretation
1	Neutral
Between 1 and 10	Limited or weak evidence in support / Slightly more probable
Between 10 and 100	Moderate evidence in support / More probable
Between 100 and 1,000	Moderately strong evidence in support / Appreciably more probable
Between 1,000 and 10,000	Strong evidence in support / Much more probable
Between 10,000 and 1,000,000	Very strong evidence in support / Far more probable
More than 1,000,000	Extremely strong evidence in support / Exceedingly more probable

One frequent difficulty in the way in which probabilistic evidence is handled in a court of law has been termed the Prosecutor's Fallacy, and relates to the Bayesian approach to uncertainty. I could use DNA testing as an example, but the very large and very small numbers involved are in themselves hard to grasp. So for the moment I'll shelve the very small probabilities associated with DNA matching and the large numbers associated with national databases and take a smaller-scale example, but one that uses equivalent logic.

A boot print is found in the flowerbed near the broken window of the burgled Bodley Manor. The print suggests a boot of middling size, but with a distinctive tread pattern. The local cobbler suggests that around one in a thousand people in the local area might have such a boot. Police make their rounds from door to door, and in the course of their enquiries they come across a pair of boots which match the print.

James Bayley, who owns the boots, is arrested. How strong is this boot print evidence? In other words, what are the chances that James Bayley is innocent, even though his boot matches the print left in the flowerbed? We know, thanks to Thomas Bayes's theorem, that this is not the same thing as the chances of Bayley's boot matching the print, if he is innocent. The probability of innocence, given the evidence, is not the same thing as the probability of the evidence, given his innocence.

Let's do the numbers. Without the boot print evidence, we should regard each one of the 10,000 residents in the neighbouring villages as a suspect. The cobbler's expert opinion is that 1 in 1,000 people would have a pair of boots that matched the print. So, of the 10,000 potential suspects, 9,999 of them are innocent people. We reckon that on average 10 would have boots that matched the print, and around 9 of those will be innocent. One person is both guilty and has a pair of boots that match the print. So the probability that an innocent person has a boot that might suggest guilt is 9 in 9,999, a little less than 1 in a 1,000. This is the probability of the evidence, given innocence.

P(Evidence | Innocence) = 0.09%.

Without the evidence of the boots, we would have no reason at all to suspect Bayley over any of the other local inhabitants; so the prior probability of Bayley's innocence is 9,999 in 10,000—so 99.99 per cent.

Plugging these numbers into Bayes's formula, we get a posterior probability that Bayley is innocent of around 90 per cent. This is the probability of his innocence, given the evidence.

P(Innocence | Evidence) = 90%.

So the evidence of the boot print alone is certainly not enough to prove his guilt. But it has made a huge difference. A 1 in 10,000 chance of guilt has turned to a 1 in 10 chance of guilt (that makes sense, since we expect that on average nine other local people might have matching boots). This is not surprising: all we have done is apply numbers to common-sense reasoning. But there are several points worth noting.

If Bayley is innocent, the probability of his boots matching the print is quite small: 0.09 per cent. We finally come to the Prosecutor's Fallacy. If Bayley was brought to trial on this evidence, the prosecutor might claim that such a small chance ('less than one in a thousand!') signals Bayley's guilt. But what Bayley's defence lawyer needs to point out is that the probability of that particular boot print appearing *at all* is very small, and that the jury should be thinking about the *relative* probabilities. The Prosecutor's Fallacy is to claim that the improbability of the evidence means evidence of improbability.

As we see from these calculations, that logic doesn't hold up. Nine or so other people from the local population could equally well have committed the crime and be faced with similar evidence. It is 90 per cent likely that Bayley is innocent, even given the supposedly damning evidence of the boot print. No judge or jury should convict him simply on the basis of the improbability of his owning boots that match the print.

The prior probability is very important in this line of reasoning. In the absence of any other evidence, we must assume that Bayley, along with every other one of the 10,000 local residents, has a 1 in 10,000 chance of being the criminal. If, though, there were any other factors that had in some way narrowed down the list of suspects, or any other positive reason to suspect Bayley, then the Bayesian formula gives a much more damning outcome. If other circumstantial evidence had established a prior probability of guilt of 1 in 10, then the addition of the boot print evidence advances the probability of guilt to 99.2 per cent, a much stronger case.

Then again, the matching boot print, while not damning on its own, is nonetheless of huge value in directing the attention of the police to possible suspects. In our story, it suggests that only 10 people are worthy of close attention.

If we substitute DNA matches for boot print matches, the numbers change, but the logic remains the same. The improbability of finding an innocent suspect's DNA at the crime scene does *not* justify the prosecutor reversing the logic and imputing the same improbability to the suspect's innocence. Modern DNA testing has advanced to the point where the chances of a random match are now extremely small. The point remains that logic requires judges and juries to compare these extremely small probabilities to the probabilities of alternative explanations. In the case of DNA evidence, the alternatives would include, for example, the risk of contamination or even malicious planting of evidence, either of which may be plausible, even if unlikely, explanations.

Crimes are unusual. Crimes like double infanticides are very unusual. Whenever a crime of this kind is investigated, we can be sure that something very improbable has in fact happened. In the case of Sally Clark, accused of murdering two babies, one of the flaws in statistical thinking was the Prosecutor's Fallacy. The proper statistical comparison would have been to contrast the chance of two tragic accidents with the chance of a double infanticide. Both of these are highly unlikely events, and the court should have compared the probabilities of the two alternatives: one attributing the deaths to chance (but supported by a more correct calculation), the other blaming a murderous mother (extremely unlikely, and admittedly a very hard thing to put a number to). That comparison between two explanations, both highly improbable, would have been unlikely to have led to a conviction beyond reasonable doubt.

The point that crimes are unusual is reflected in the legal principle of presumption of innocence: an accused person is innocent until proven guilty. We can cast this in Bayesian terms: in the absence of evidence, the prior probability of guilt is very low. It's the chance that someone selected at random off the street is in fact the person responsible for the crime being tried. For the posterior probability to rise to a level that is 'beyond reasonable doubt', strong supporting evidence must be presented, and the more of it, the better.

Priors and prejudice

It's easy to draw conclusions about individuals from statistics about groups. Imagine a city with two rival football teams, the Yellows and the Greens, who have a history of confrontation. The police keep records of trouble-makers, and these records show that supporters of the Yellow football team start more fights than the fans of the Greens, by a margin of 3 to 1. You're a magistrate, and are asked to decide a case involving a dust-up between two football fans. You skim the case notes and learn nothing other than that they support different teams: one is a Yellow and the other a Green. Who do you think is more likely to have started the brawl? You know the relevant statistics, and you may be inclined to think that the one in the Yellow shirt is the trouble-maker. Would that be justified? Should you find a person guilty on the basis of that evidence alone?

In Bayesian terms, you might say that there is a prior probability of 75 per cent that the Yellow supporter is to blame. But there is a trap here. If you are tempted to apply a group statistic derived from a broad analysis to a more narrow purpose, you run the risk of making an unfair judgement. This is an example of what's termed an ecological fallacy: when a statistical proportion derived from a wider population is applied as a probability to an individual or a smaller group. It is premature and unfair to make a judgement about an individual case based only on information that relates to a group to which they belong.

Now the two fans are brought in: the one in the Green shirt is a strapping, muscular lad, full of bluster and self-importance: the one wearing Yellow is a slight, timid-looking individual. Does your presumption of guilt change? You need proof beyond reasonable doubt. You'd better start finding some real evidence, for you've realized that to make a decision based on the colour of the shirt was wrong, and that it would be equally wrong to judge guilt or otherwise based on physical appearance. After all, there is a word that means to have come to a decision in advance, to have pre-judged the matter. That word is 'prejudiced'.

CHAPTER 15

Clarity from Chaos

A statistical analysis, properly conducted, is a delicate dissection of uncertainties, a surgery of suppositions.

M. J. Moroney

Rest in peace

In the county of Statfordshire, the town of Little Stirring has a reputation as a quiet, affluent location. Its neighbour Livebury, with the same population size (50,000 people), is very different: a buzzing, commercially focused market town. The citizens of Little Stirring, though, enjoy a higher average income and a high standard of living.

But when Kay, choosing where to buy a home, stumbled across some statistics, she was very surprised to find that the death rate in Little Stirring was more than double that of Livebury. Having something of an inquisitive nature, she decided to delve into the detail a little. These are the numbers she found:

Age	Little Stirring			Livebury		
	Population	Deaths	Death rate / 10,000	Population	Deaths	Death rate / 10,000
0 to 19	5,000	1	2	15,000	6	4
20 to 39	8,000	12	15	15,000	30	20
40 to 59	9,000	72	80	10,000	90	90
60 to 79	18,000	900	500	7,000	490	700
80 to 100	10,000	5,000	5,000	3,000	1,800	6,000
All ages	50,000	5,985	1,197	50,000	2,416	483

What are the Chances of That? How to think about uncertainty. Andrew C. A. Elliott,
Oxford University Press. © Andrew C. A. Elliott 2021. DOI: 10.1093/oso/9780198869023.003.0015

She was astonished to see that in every age band, the death rate for Little Stirring was lower than that for Livebury, and yet taking all ages together, her original information was confirmed: Little Stirring's death rate was more than twice that of Livebury.

How could this be? Answer at the end of the chapter.

Galileo

Galileo Galilei (1564–1642), born in Pisa, is recognized as one of the most influential scientists to have lived, and he fits many of our ideas of what a scientist should be like. He made observations (the moons of Jupiter, the rings of Saturn), he thought about what those observations might mean, and he made theories to explain his observations. He was also by all reports a great experimentalist and teacher, bringing his theories to life using dramatic images and (reportedly) performing actual demonstrations such as the dropping of cannonballs from the top of Pisa's leaning tower. He was challenged by the Church over his conviction that the Earth orbits the Sun, and even though he was forced to publicly recant this theory, the story goes that under his breath he muttered the words 'and yet it moves'. The strength of his conviction came from observation and reasoning, even if this went against the authority of the Church.

One of Galileo's chief interests was to understand how things fall. Without slow-motion cameras, it is hard to make precise observations of a falling object.

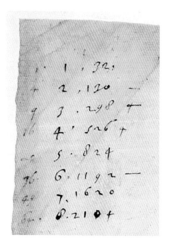

But Galileo found a way to create his own slow motion, to properly study the movement of an object pulled by the force of gravity: he conducted a series of experiments investigating the movement of a ball rolling down a gently sloping surface. In this way he effectively weakened the force of gravity. In the absence of modern timepieces, he made ingenious use of a water clock to measure time.

For one of his experiments, he noted how far the rolling ball had progressed for every unit of time that had passed. His notebooks show a record of this experiment. He wrote down how many 'points' of distance a rolling ball had covered for every unit of time (we'll call it a 'tick') that had elapsed. This is what he found.

Time	1 tick	2 ticks	3 ticks	4 ticks	5 ticks	6 ticks	7 ticks	8 ticks
Points	32	130	298	526	824	1,192	1,620	2,104
We can play around a little with his numbers:								
% of total	1.52%	6.18%	14.16%	**25.00%**	39.16%	56.65%	77.00%	100%
Re-scaled	1	4.06	9.31	16.4	25.75	37.25	50.63	65.75
Close to	1	4	9	16	25	36	49	64

Inspection of these numbers shows several things. For the first clue, look at that 25 per cent near the centre of the table. This tells us that in half the time that the experiment took (four ticks), the ball had rolled a quarter of the total distance. There is a clue here: a half, squared, equals a quarter. Could it be that the distance travelled is related to the square of the time that has passed?

We can go further and express those distances using a different unit, one derived from the data itself. If we treat the distance travelled in the first tick of time as one unit of distance, then we get the row labelled 'Re-scaled'. It's not too hard to see that these numbers are very close to the squares of the first eight counting numbers. Roughly speaking, after one tick, the ball has rolled one unit of distance; after two ticks it has rolled four units; after three ticks, nine units, and so on, until after 8 ticks the ball has rolled about 64 units of distance. Look again at Galileo's notes: in pale script to the left of his numbers 1 to 8 are the squares of those numbers: 1, 4, 9, 16, 25, 36, 49, 64. There's no doubt that he had spotted the mathematical relationship: the distance rolled is proportional to the square of the time taken.

This may not seem so astonishing: we are used to the idea that mathematical laws can describe the physical world, but it is truly a profound insight. Galileo had built an experimental set-up, had made measurements as precisely as he could, and, interpreting the numbers, he had established a relationship between measurements in the physical world and a theoretical, ideal model expressed using the abstractions of mathematics. Mathematics was embedded in reality. In his words: 'the book of nature is written in the language of mathematics'.

There were, of course, more questions to be answered: was this 'law' that was true for balls rolling down inclined planes also true for an object falling freely?

(It was, allowing for gravity operating at full strength.) Would it be true if the object was not initially at rest but was already moving? (It would not, but the adjustments to the model to allow for an initial velocity were straightforward.)

Isaac Newton would later provide a deeper explanation of these laws of motion, adding his understanding of gravity and using his newly invented mathematics of calculus to show how a constant force accelerates a falling object and to demonstrate mathematically that the distance covered *must* be mathematically related to the square of the time elapsed. That's how it goes: an experiment reveals a pattern in the observations; the pattern can be described mathematically; the mathematics leads to a deeper explanation. Galileo found the model; Newton explained it.

But there is another lesson here: Galileo's measurements were imperfect, and the numbers do not *exactly* match the mathematics. There are some small discrepancies: something else to be explained. Galileo was working with very rough-and-ready measuring equipment; so it's reasonable to think that the discrepancies may be due to imperfect measurement. We can think of his measurements (32, 130, 298, etc.) as being composed of two parts: the true, idealized value and a measurement error. The time-squared rule he detected was the signal, the measurement error was noise. His mathematical formula (his 'model') accounted almost but not entirely for the values he recorded. The amounts unaccounted for are small, and (if we take the trouble to fit a formula to his values a little more carefully) are seemingly without obvious pattern. 0, −0.1, 0.03, −0.1, 0, 0.1, 0.1, −0.2. Statisticians call these unexplained fragments 'residuals', and in a more extensive experiment with more data we might try to discover whether or not there was any further pattern to be found in the residuals.

In Galileo's case, the signal of the squares came through loud and clear and the noise of error was the gentlest of hisses. So Galileo discovered the formula for a ball rolling down a slope. What about other bodies in motion? He thought about cannonballs and found that their trajectories were approximations to parabolas.[147] But if we do actual experiments with real cannonballs, we find that the measurements are not so perfect: the errors no longer look random. For gravity is not the only force involved: there's air resistance, there's the wind blowing across the field. This doesn't make Galileo wrong about the effect of gravity over time and distance, it simply means we need to add additional elements to account for the new, smaller forces.

---

[147] A parabola is the shape of the graph of a square function like $y = x^2$. On the page of Galileo's notebook preceding the one shown, he has sketched a parabola.

The truths that we wrest from observations are always provisional and incomplete, and even in the best of cases they always contain a grain of uncertainty, a little room for improvement. But we can develop better and better models that make better and better predictions of what to expect in this random world.

Meanwhile, back at CERN

Let's drill a little deeper.[148] CERN is the European Organization for Nuclear Research and is based in Geneva, on the border between Switzerland and France. The high-profile experiments at CERN are centred around the Large Hadron Collider (LHC), a 27-kilometre-long circular tunnel ('large'), in which particles, typically protons ('hadrons'), are accelerated in two beams to almost the speed of light and then smashed together (it is, after all, a 'collider'). The aim of this kind of high-energy physics is to find unusual particles, typically heavy particles. $E = mc^2$, as Einstein deduced; mass and energy are two aspects of the same thing, and if you have enough energy involved in a collision, the energy can turn into mass in the form of very short-lived particles. In the case of the LHC, a main goal of this impressive piece of experimental equipment was to find the Higgs boson, a particle whose existence had been predicted (by Peter Higgs and others) in the 1960s. Their model, their theoretical understanding of the physical world, had a space for a particle with specific characteristics, but one which had never been observed, and could only be observed in the sort of conditions that the LHC was designed to produce. The missing particle would be much heavier than the proton (perhaps up to 1,000 times heavier), and that needed a high-energy experiment. Einstein showed that at relativistic speeds—close to the speed of light—the mass of particles increases, and in an apparatus like the LHC, the protons would acquire sufficient mass and energy to allow particles like the Higgs boson to form. The experiment is exquisitely controlled, bringing the two beams together with great precision, but when they collide, things get chaotic—particles of all sorts emerge from the high-energy collisions. It is the job of the various detectors around the LHC experiments to spot, and to make measurements, of the particles that fly out, notably of their energy. And it is the job of the physicists to make sense of the flying particles and interpret their trajectories.

---

[148] How deep? The LHC tunnel is around 100 metres below ground.

The picture that emerges is a statistical one. There is a distribution of particles at different energy levels, and some level of random departure from a theoretical background baseline is anticipated. If nothing of interest is happening, this follows an expected statistical distribution. About one-sixth of the time, cases fall more than one standard deviation (one sigma) above the expected value,[149] and about 2.5 per cent of the time, beyond two sigmas. And outside five sigmas? That would happen by chance 0.000029 per cent of the time, or about one in 3.5 million times.

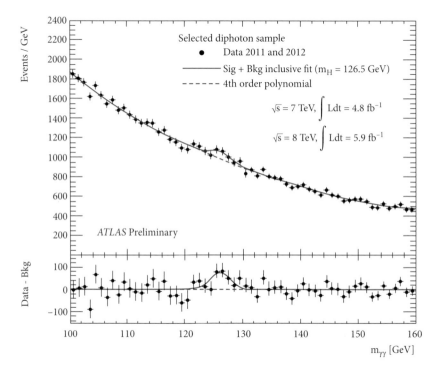

This graph from the ATLAS experiment shows how many events occur at each mass–energy level. If there were no Higgs boson, the bump in the middle at around 125 GeV, five sigmas from what was expected, would have had to have arisen by chance. The statistical calculations say that this would have been at odds of 3.5 million to 1. And this was not the only evidence supporting

---

[149] The rule of thumb is often given as *one-third* of the cases fall outside one standard deviation. That is the case in a two-tail test where we are as interested in values that are very low as in those that are very high. In this case, though, the physicists were interested only in high values; so only the upper tail of the distribution is used.

the Higgs discovery: a second experiment (CMS) at the LHC had independently found a bump at around the same mass–energy level, also at the five-sigma level. With two independent experiments reporting results that were so unlikely to have arisen by chance, the scientists felt justified in breaking out the champagne.

The scientists at CERN may be physicists, but they too need an understanding of chance and uncertainty. They use the ideas and language of probability and distributions to measure and express the degree of confidence that they have in their findings, as their ideas progress from suspicions to damn-near-certain.

Asking *why?*

Children are natural scientists: feeling, tasting, testing everything around them. They gather empirical evidence, they notice patterns and make observations that allow them to form models of the world from which they then make useful predictions. We all learn to judge distance, time, weight, temperature using those highly approximate measuring devices, our bodies. We make mistakes, and sometimes hurt ourselves, and from this we learn. We are all casual statisticians, making observations and weighing up what lessons we can draw from the confusing hubbub around us.

And as children do, we ask *why?* We want to see inside the box, and why things work the way they do. We can never directly observe the laws of nature: all we can do is make better and better explanations. Often even our best explanations are incomplete and approximate, and so we look for better ones, gathering more information to try to build better models to tell us what is in the box.

Correlation? Causation?

Correlation doesn't imply causation, but it does waggle its eyebrows suggestively and gesture furtively while mouthing 'look over there'.

Randall Monroe, *XKCD.com*

It's said that one of the first things students of statistics learn, and one of the first things they forget to apply in practice, is: 'Correlation does not imply causation'. It's not hard to find examples of meaningless statistical associations. Tyler

Vigen's website *Spurious Correlations* hosts numerous data sets and hunts out strong but wrong correlations. Try this one:

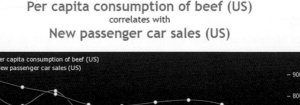

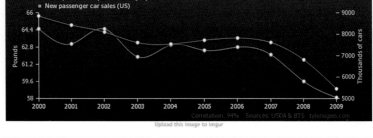

	2000	2001	2002	2003	2004	2005	2006	2007	2008	2009
Per capita consumption of beef (US) Pounds (USDA)	64.5	63.1	64.5	61.9	63	62.5	62.8	62.1	59.6	58.1
New passenger car sales (US) Thousands of cars (BTS)	8,847	8,423	8,103	7,610	7,545	7,720	7,821	7,618	6,813	5,456

Correlation: 0.938256

A correlation of 1 means that two data sets were moving in lockstep, that you could perfectly predict a change in one from a change in the other. This example shows a correlation of 0.938 between beef-eating and car-buying. And yes, these two data sets, over the years 2000 to 2009, moved in very similar ways, but don't think that means much. That website is specifically set up to commit the crime of analysis in hindsight—painting the target after the sharpshooter has fired his shots—and it will trawl every possible association before presenting the one that best fits the sought-after narrative. If there is a shred of meaning here, it is that both beef consumption and new car sales were declining over the period covered, and have a common association with time passing. It certainly doesn't imply that car sales would be boosted if only people would eat more beef (or that people would stop eating beef if only they could stop buying cars).

One of the problems in establishing that one thing causes another is the possibility that both are influenced by some third factor (in the example above, the passage of time seems to have had some effect on both car sales and beef consumption). Researchers go to great lengths, when designing experiments, to account for these possible confounders. One of the virtues of randomized control trials, the so-called 'gold standard' of experimental design, where subjects are allocated at random to test groups, is that they cope with the effect of confounding factors, including those that cannot be easily measured, and even those that are unknown. In effect, by randomizing the test subjects, the experimenter introduces a little noise to mask the unwanted signal of possible confounders.

So we cannot fault the statistical orthodoxy that correlation does not imply causation. Here, though, is the often unstated problem: causation is what we really want. Reluctantly, we accept that mere correlation can't give us this. But cause and effect is what's needed. Want to limit heart disease? Well then, are there lifestyle changes that would help? Gun violence a problem? What could we do to make a difference? Why do so many people die in motor accidents? Is it due to driver behaviour? Car design? Poor road signage? All of these to varying degrees? These are the questions to which we really want the answers, and these are the questions that all budding statisticians learn cannot be answered simply by correlating data sets.

In the twentieth century, one life-and-death question of this kind sparked debate, protest, and lawsuits over decades. Honest investigators, biased manufacturers with vested interests, and confused consumers were all enveloped in a cloud of statistical information and misinformation as they tried to arrive at a definitive answer. The question was: what lay behind the dramatic increase in lung cancers in the first part of the twentieth century?

Does cigarette smoking cause lung cancer?

The harmful effects of cigarette smoking are now so well established and so broadly accepted that it seems strange even to ask that question, but 70 years ago this was by no means clear. Looking back at that time from our privileged, filtered-by-hindsight viewpoint, we can note two parallel trends: two phenomena that increased from near-zero levels, and grew in similar ways, albeit separated by a period of around 25 years.

The first was cigarette smoking. A cigarette-rolling machine was invented in 1881 by one James Bonsack, and the product, which was cheap and easily distributed, quickly became very popular. The second trend, kicking in about a quarter of a century later, was an unprecedented growth in the incidence of lung cancer.

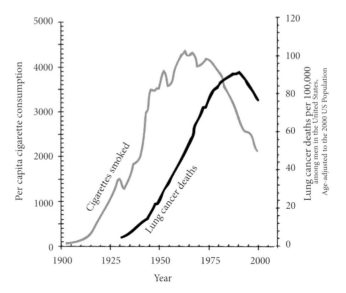

To us, looking at this graph with the benefit of hindsight, the message seems very obvious: cigarette smoking causes lung cancer, and it takes about 25 years to do so. The upward-sloping lines tell a powerful tale: the more smoking, the more lung cancer. Even more persuasive is the topping-out and decline of the curves: reduce smoking, and 25 years later, lung cancer reduces. There are few now who would doubt that this is true, but at the time it was by no means a clear-cut argument.

Hindsight and **foresight** tell different stories. An article written in *The Atlantic* magazine in 1956[150] about smoking and lung cancer makes this clear. It gives us the flavour of mainstream thinking, at a time when opinion was not yet settled, but when the first strong statistical evidence was emerging. The article notes the then-recent dramatic rise in rates of lung cancer: a disease that had been very rare indeed in 1900 had now become common. In the period from 1930 to 1948 the death rates from lung cancer in men in the United States increased from 5.3 per 100,000 to 27.1 per 100,000—a five-fold increase. Whatever the causative factor was, exploratory statistics suggested that it had appeared recently, was more common in males than females, was more common in the city than in rural areas, and was likely to be due to a substance that was inhaled.

---

[150] 'Lung Cancer and Smoking: What We Really Know'. C. S. Cameron, *The Atlantic*, January 1956.

There was more than one candidate for the causative agent. The first half of the twentieth century was a time of great change. It was not just cigarettes that were a novelty. The list of suspect substances included industrial fumes, soots, exhausts from internal combustion engines, fumes from bituminous road surfaces, and cigarette smoke. Plausible cases could be made for each of these. The article in *The Atlantic* concludes that smoking cigarettes is the prime suspect, on the basis of evidence that was then emerging from innovative research in both the United States and the United Kingdom. In the words of the author in 1956, 'If it has not been proved that tobacco is guilty of causing cancer of the lung, it has certainly been shown to have been on the scene of the crime.'

At the end of the Second World War, the United Kingdom had the highest incidence of lung cancer in the world. Physician Richard Doll, doing research into asthma at St Thomas's Hospital in London, initially thought that the cause might be car exhausts or the fumes from tarmacking roads. Then in 1950 Doll, then aged 28 and working with Austin Bradford Hill,[151] undertook a study to look at lung cancer patients in 20 London hospitals.

We ask if cigarette smoking causes lung cancer, but that question can be answered in two ways, forwards and backwards. Smoking a single cigarette quite clearly does not invariably give you lung cancer. It is a matter of chance, and it is a matter of degree. For each cigarette smoked, the risk of getting cancer accumulates by some small amount. We all know lifelong smokers who seem to defy the odds. It's not even the case that *most* smokers get lung cancer: relatively few of them do. In the statistics quoted by the 1956 article in *The Atlantic*, it is suggested that the lifetime risk of a young man who smoked a pack a day getting lung cancer was around 1 in 15 to 20. But this is set against the rate for non-smokers: just 1 in 170 to 190. Smoking led to around a tenfold increase in risk, but by no means to the level of inevitability.

But the question that Doll and Hill asked in their initial study was a subtly different one: is there some factor that links those with lung cancer? The evidence was clear and persuasive. The outstanding common element in their 1950 study was that those with lung cancer had, to an overwhelming degree, been smokers. Doll himself gave up smoking immediately. In their 1950 paper, Doll and Hill wrote:

---

[151] Hill was the statistician behind what is considered the first randomized controlled trial in 1946: an investigation into the use of streptomycin for the treatment of tuberculosis.

'The risk of developing the disease increases in proportion to the amount smoked. It may be 50 times as great among those who smoke 25 or more cigarettes a day as among non-smokers.'

So not all smokers get lung cancer, but on reversing the direction of the logic, they found that almost all lung cancer patients were smokers. What was causing the rise in lung cancers? Doll and Hill were sure that the clear answer was cigarette smoking. But it was the work of decades to gather the evidence that would make the case beyond argument.

In 1951 Doll and Hill launched a more ambitious project. This was to be a prospective study, forward-looking, tracking a 'cohort' of subjects over a long period. And they looked not just at lung cancer but also at other diseases, aiming to assess the overall impact of smoking on health over life. As subjects, they enlisted 40,000 UK doctors, judging that follow-up contacts would be easier in a population where people would be motivated towards ongoing participation. Since the doctors were registered with professional bodies, this would also make it easier to keep track of them. As it turned out, they were able to keep the study going for a remarkable 50 years, finally closing the books in 2001. The results were impressive. These survival curves show the percentage of participants who survived to various ages.

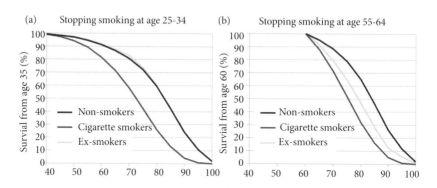

As shown, lifelong cigarette smoking was found to be associated with around 10 years' reduction of lifespan. Remarkably, the British Doctors Study also found that stopping smoking was unquestionably worthwhile. For someone stopping smoking around age 30, as shown in the first graph, their future survival rate would be very similar to that of non-smokers, while stopping smoking even as late as age 60 resulted in substantially improved survival rates.

Dozens of other studies have since produced similar results, and few would now argue that the link between smoking and lung cancer is mere correlation.

But even now, 70 years after Doll and Hill's first study, smoking is still common in many parts of the world. Opposition to anti-smoking campaigns from tobacco companies has been relentless.

Responsible scientists who understand uncertainty know that even the most robust statistical analysis leaves a gap for being wrong, for doubt. We can seldom be 100 per cent certain of anything, but falling short of absolute certainty is no excuse for inaction. Those arguing against anti-smoking regulations, though, have latched on to any small element of doubt, saying 'but you can't be absolutely sure', and still continue to peddle their poisons.

Seeking the cause

Many scientists have been traumatized to learn that none of the methods they learned in statistics is sufficient to articulate, let alone answer, a simple question like 'What happens if we double the price?'

Judea Pearl

Few would now dispute that evidence justifies the claim that the leading cause of lung cancer over the last century has been cigarette smoking. One of Richard Doll's more prominent statistical opponents in the debate over smoking and lung cancer was the statistician Ronald Fisher. Fisher was a formidable statistician: he had been responsible for creating much of the toolkit of modern mathematical statistics. He did not dispute the *possibility* that smoking was a cause of cancer, but he vehemently opposed the notion that the case had been *proved*. He himself was a lifelong smoker, and he had acted as a consultant to the tobacco industry. Fisher accepted the correlation between cigarette smoking and lung cancer, but argued that the correlation could be the consequence of some other mechanism. Perhaps pre-cancerous changes in the lung caused an irritation that could be soothed by smoking; perhaps there was a hidden confounding factor—some genetic predisposition that led both to cancer and an urge to smoke.

The debate about causality in regard to disease was not a new one. Robert Koch (the same Koch who grew bacterial cultures on potato slices), working with Friedrich Loeffler, had in 1884 formulated four 'postulates' that he felt were fundamental to claiming that a microorganism (a germ) was the cause of a disease:

- The microorganism must be found in quantity where the disease was present, but not where it was absent.
- It must be possible to isolate the microorganism and grow it as a pure culture.
- The cultured microorganism, when introduced to a healthy host, must result in the disease.
- If then isolated from the deliberately infected host, it must be identical to the original culture.

Koch's postulates are mainly of historical interest. They specifically apply to microorganisms that can be cultured and not, for example, to chemicals in the environment. Even so, there are many exceptions to his postulates. Nonetheless, they represent an early attempt to formalize the question of causation. Richard Doll's mentor, Austin Bradford Hill, took the matter further and listed his own rules for claiming something as the cause of a disease. He identified nine criteria:

- **Strength**: The association between cause and effect is strong.
- **Consistency**: The cause–effect link is observed in many different contexts.
- **Specificity**: There exists a one-to-one relationship between cause and effect.
- **Temporality**: The effect comes after the cause.
- **Biological gradient**: More of the cause leads to more of the effect.
- **Plausibility**: There is a plausible mechanism whereby the cause can lead to the effect.
- **Coherence**: The overall phenomenon of cause leading to effect makes logical sense.
- **Experiment**: There is experimental evidence to support the cause leading to the effect.
- **Analogy**: The link between cause and effect resembles other known cause-and-effect relationships.

Hill's criteria are not mathematical in nature: they are more like a codification of common sense. They need not all be present to make a convincing case for cause and effect, so the notion of proof that they support is nearer to that found in a court of law than in a mathematical theorem. And, as decades of lobbying for the tobacco industry has shown, they are vulnerable to attack with FUD— fear, uncertainty, and doubt. Because they are not mathematically watertight, it's easy enough to lazily say: 'but you can't be sure of that'.

More recently, new mathematical approaches to establishing causation have been developed, by Judea Pearl and others. One of the key ideas behind this work is to move from a passive version of a conditional probability to a more active one. A conventional statistical approach might calculate the probability of a heart attack, given a patient's exercise regime. The causality statisticians wonder about the probability of a heart attack if the patient is persuaded to make a change in their exercise regime. They make a distinction between writing $P(A|B)$ and writing $P(A|do(B))$, the probability of outcome A, given that you **do** intervention B. Careful analysis of diagrams of possible cause-and-effect relationships allows a clearer understanding of what variables might or might not confound an analysis, and how to control for them, and in the end to tease out the strength of causal connections from experimental observations. This approach can bring us closer to the kinds of questions we really want answers to: What can we do to make things better? What happens if we exercise more? What happens if we stop burning fossil fuels? And even, to answer Judea Pearl's question, what happens if we double the price?

Answer to:
Rest in peace

The numbers are not wrong, and nor is the arithmetic. In every age group, the death rate in Little Stirring is lower than that in Livebury. And yet, taking each town as a whole, Livebury has a lower death rate than Little Stirring. This is an example of Simpson's paradox. When you look at the big picture (large groups of subjects), it seems that you see one effect, but when you break the large groups down to smaller subgroups and investigate those, the effect reverses.

One way of thinking about this is that in our example, there are two main sets of influences on the death rate that account for differences between the locations: there are those that have to do with age, and everything else. The residents of Little Stirring are older, and so more of them die each year. But on the whole, it appears to be a healthier place to live. If you look at groups of equal age, then the age effect disappears and other factors come through. Then we see that, for whatever reason, Livebury residents experience a higher death rate in every age band.

The moral of this story is to consider carefully what it is you are trying to investigate and try to accommodate the effect of any factors that may interfere with that analysis. In the case of Kay, age is a given, not a variable, and should, if possible, be removed from the analysis.

CHAPTER 16

Sharing the Risk

Insurance by its very principles is collection of money by many to pay to a few, in their times of distress. Rather than considering it to be a passive investment, it should be treated as an active social participation where you are securing yourself as well as helping others in their direst times of need.

Tapan Singhel

What's the worst that could happen?

You lift your glass in silent greeting to the elegant stranger as he enters the hotel bar and comes over to stand alongside you. He orders a mineral water (with a twist of lemon) and takes a sip.

You break the silence: 'What, no tempting bet for me tonight?'

He pauses. 'If you'd like one,' he replies.

'Go on, then.'

A moment's reflection followed by a quiet chuckle: 'Okay, try this...'

He reaches into his pocket and pulls out a roll of dollar bills. He peels off eight and pushes them along the counter towards you.

'Here's the deal. You're the banker tonight. That's your starting capital. I'll flip a coin. Heads I win, and you pay me a single dollar. You keep the rest of your capital, seven dollars. But if it's tails, I flip again for twice the amount. So then, if the coin comes up heads, I win two dollars. You keep six. And so it will continue. As long as it keeps coming up tails, I keep flipping, with the stakes doubling each time. So if I win only on the fourth flip, you give me all eight dollars back, and we're square.'

What are the Chances of That? How to think about uncertainty. Andrew C. A. Elliott,
Oxford University Press. © Andrew C. A. Elliott 2021. DOI: 10.1093/oso/9780198869023.003.0016

'And if you win only on the fifth go?'

'Well, that's unlikely, but if it happens, then you pay me 16 dollars—the eight I have staked you, and eight of your own. If I lose on that roll, we still keep going, doubling each time.'

You think this through. There's a 50/50 chance that he will succeed on the first flip and you'll only have to pay out one dollar. If he gets tails on the first try, then there are still two more chances for you to end up with a profit. For you to actually lose money, he must flip tails four times in a row. Only a 1 in 16 chance of that. It's tempting. Finally you respond.

'No. Though I am very likely to win, I could lose, and then I lose quite a lot. So no. Not tonight; not this one.'

'I see you're learning,' he smiles. 'Eight dollars of starting capital is not enough for you. Tell me, how much would I have to stake you to play this game with me?'

Well, what's your answer? Discussion at the end of this chapter.

Life insurance for astronauts

In 1969, before the launch of the *Apollo 11* mission that would take Neil Armstrong and his fellow astronauts—Buzz Aldrin and Michael Collins—to the moon and back, Armstrong estimated that the chance of mission success was around 50 per cent. Unsurprisingly, the crew thought they should take out life insurance. But there were a number of hurdles. Only a few insurers were prepared to contemplate taking on the risk, and the premiums that they were asking were exorbitant—way beyond the reach of those on a NASA pay scale, however prominent they might be. NASA themselves were prohibited by law from taking out insurance. One insurance company did offer to arrange free cover for them, but there was a condition: publicity for the company and their sponsors who would share the cost of the premiums. Armstrong was unwilling to agree to these terms.

So a plan was hatched. While the astronauts were in pre-launch quarantine, they each autographed hundreds of stamped postcards—Michael Collins estimates that they signed 1,000 of them. These covers, as they are called by philatelists, were produced in collaboration with the Manned Spacecraft Center Stamp Club in Houston. Once the covers had been signed, they were bundled up and entrusted to a collaborator, who had instructions to take a batch of them to the Post Office at the Kennedy Space Center on the day of the launch, and there to ensure that the stamps were cancelled with a postmark bearing the date—16 July 1969. The remaining batches were returned to Houston and cancelled with a Houston postmark on the day of the moon walk. Finally, the covers were given to the families of the crew for safekeeping. Should the mission fail, the reasoning went, the insurance covers, as they became known, would become highly valuable and provide their families with some form of financial assistance.

There never was a need to make a claim against that unusual insurance arrangement, but those covers are now highly sought-after collectibles, selling for tens of thousands of dollars.

Safety in numbers

Paying for insurance is in some ways like playing the lottery. You repeatedly pay in a relatively small amount, and you do this in the expectation that from time to time you may receive a large payout. The difference is that in the case of insurance, the payout is explicitly compensating for a loss, a chance event that will cause material financial damage, be it death, illness, fire, flood, motor vehicle accident, or a broken toaster. If buying insurance is somewhat like gambling, then the insurance company plays the role of the casino. And just as a casino can exploit the law of large numbers to create a reliably profitable business based on the operation of pure chance, so too are insurance companies risk-averse participants in a business that is based on risk. Their security comes through providing cover to very many policyholders and covering very many risks. If an insurer manages its affairs well, the random occurrence of claims will be smoothed away by the big numbers involved and they will remain financially stable, despite the fact that chance is at the heart of their business. The risks they are covering are pooled and, despite the randomness of the incidence and size of the claims that must be paid from it, if the pool is large enough, it will not itself be at significant risk. If the actuaries and underwriters have done their job, the fluctuations will be manageable. Individuals will be protected and the collective reserves will be large enough to bear the random variability.

The principle behind insurance is closely related to the **individual–collective** duality. The deal offered to you, the individual buyer of insurance, is the opportunity to share in this collective averaged pool. Instead of having to ensure that you have sufficient reserves to weather any storm, you can draw on the insurer's reserves, which can be smaller proportionately—thanks to the pooling effect. Instead of suffering the financial shocks of random misfortunes, you make a payment that reflects not the full cost of those unwelcome events, but only their probability.

The premium you pay for insurance to cover that risk is based on the average cost of claims that you and those like you will make (plus something extra to cover the insurer's costs and, of course, some profit). The insurance bargain is that you are trading your individual vulnerability and exposure to outlier events for membership of a group that gives you collective security.

Insurance companies have something of a negative reputation. Pushy insurance salespeople, fine print that allows claims to be refused, financial muscle that is sometimes used in questionable ways, and a reluctance to take on certain risks are all practices that have contributed to an unlovely and unloved stereotype. But insurance is rooted in worthy principles: it grows from strong community values and the need for cooperation and mutual support.

A friendlier society

As a student in Cape Town, an annual event I attended was the Community Carnival at Maynardville, a festival organized by the Rotary service organization. Bands would perform, there were carnival games to play, and various immigrant communities would host stalls offering typical national foods, from the akvavit provided by the Norwegians to the pickled herring available at the Dutch stall. The most convenient public parking was a few blocks from the park entrance and on the walk from the car we would pass a quirky-looking building with the intriguing title 'Oddfellows' Hall'. What I didn't know then is that this building had something in common with the community spirit of the Rotary Club's annual carnival.

People with shared interests have always recognized that gathering together for mutual defence is an effective strategy for softening the blows of ill fortune. In England in the eighteenth century, with the decline of trade guilds, craft and tradespeople who were otherwise unaffiliated with larger organizations started forming associations for mutual support. These associations were 'Friendly Societies', and several of them referred to their members as 'Oddfellows'. While I can find no definitive explanation of this intriguing name, the suggestion I find most plausible is that these fellows were odd not in the sense of being peculiar, but in the sense of being freelancers, self-employed, unattached to, and so not protected by, larger guilds or associations.

Oddfellows' Hall, Wynberg, Cape Town (2010).[152]

In addition to the social functions offered by these societies, they started to act as providers of financial security to the members. In return for a monthly contribution, a member would be entitled to a modest amount of sick pay, a death benefit, and in some cases a pension.

This mutuality movement gave rise to many variations: funeral societies, credit associations, mutual savings banks, even service organizations like Rotary, all with a similar aim of protecting and supporting the position of vulnerable individuals through the sheltering effect of affiliation with a larger group.

As the friendly society movement, including Oddfellows' societies, grew in the United Kingdom,[153] they were recognized in law as Friendly Societies, providers of insurance and other services, but operating under different laws to those governing insurance companies. Even today, active Friendly Societies operate under the 1974 law, although no new societies may be registered under that law. One of these societies still trades under the name The Oddfellows[154]— with over 300,000 members, they offer financial services, but also emphasize their ongoing commitment to their fraternal traditions.

---

[152] Picture provided by kind permission of David Brebner. Recently, the Wynberg Oddfellows' Hall has become an Open Mosque.

[153] And not just in the United Kingdom: Oddfellows' Halls can be found throughout the world.

[154] Formally, though, they are The Independent Order of Oddfellows Manchester Unity Friendly Society Limited.

Many larger insurers were established under rules of mutual ownership. My first employer was the South African Mutual Life Assurance Society, better known nowadays as the Old Mutual. Mutual companies were technically collectively owned by the members, but in practice they operated much like straightforward shareholder-owned companies, the members having little real say in decision-making. Mutual ownership places restraints on corporate growth, however: there is no easy way to attract new capital to fund expansion, and in the last part of the twentieth century many insurers abandoned the principle of mutuality and became shareholder-owned.

Scanning a list of insurers registered in the United Kingdom shows that there are some whose original location, purpose, and values are still remembered in their names. A few examples: Scottish Widows, Assured Guaranty, Carolgate Friendly Society, Dentists' and General Mutual Benefit Society, Exeter Friendly Society, The Prudential, Steadfast Friendly Society, Travelers Insurance Company. There's little room for nostalgic sentiment in modern business, and insurance is no exception: taking out an insurance policy is a commercial transaction, where each party seeks the best deal. Still, these names remind us of their origins in mutual support and cooperation.

For those in peril on the sea

Small insurance policies are seldom worthwhile. The extended warranty on your toaster is unlikely to offer value for money. Better to save the cost and buy a new one if it fails. But where risks are substantial, where the potential loss could mean ruin, insurance makes sense. And few risks are more substantial than that of losing a ship at sea.

When things go to plan, shipping is a very effective way of moving goods and people. But there are risks. A ship lost is a disaster. As mentioned earlier in the chapter 'Reckoning the Chances' (p. 44), the ancient Greeks and Romans recognized the perils involved and they reflected this in the way they did business. Investors would make a so-called loan on bottomry to fund a trading voyage. This incorporated a kind of insurance: in the event of a total loss of the vessel, the loan would not need to be repaid. The investors bore the risk, and in return they asked for a greater return. In ancient Rome a typical premium for this kind of risk was thought to be around 12 per cent, although this premium was never separately identified. Similar contracts were available to traders in twelfth-century Italy, and through later centuries, the practice spread throughout Europe.

By the seventeenth century, such contracts were common in England. In the forty-third year of the reign of Queen Elizabeth I (1601), the Policies of Assurance Act regarding marine insurance was passed. The preamble to this legislation describes this practice of making specialized loans for voyages:

> by meanes of whiche Policies of Assurance it comethe to passe that upon the losse or perishinge of any ship there followethe not the undoinge of any Man, but the losse lightethed rather *easily upon many* than *heavily upon fewe*, and rather upon them that adventure not, than upon those that doe adventure, whereby all merchants, spiallie the yonger sort, are allured to venture more willinglie and more freelie. (My emphasis)

I think the principle of insurance has seldom been better expressed: a loss alighting easily upon many, rather than heavily upon few.

Lloyd's of London

In the *London Gazette* No. 2429, for the third week of February in 1688, there appeared an advertisement offering a reward of one guinea for information relating to the theft of five watches. To claim the reward, the informer was directed to approach 'Mr Edward Lloyd at his Coffee House in Tower-Street'. This is the first recorded mention of Edward Lloyd's famous coffee house, which would go on to become the remarkable institution that is Lloyd's of London, a huge and singular insurance market.[155]

Edward Lloyd's establishment was a popular gathering place for merchants and seamen, something that Lloyd encouraged by making available as much information as he could regarding ships and shipping. He organized news-gathering channels to track the movements of sea vessels all over the world, and so by becoming a hub for marine intelligence, Lloyd's also became a place for related business. Merchants and underwriters based themselves there as a convenient place for arranging marine insurance. As Lloyd's changed from coffee house to insurance market, it went from strength to strength, relocating to ever more prestigious locations.

In October 1799, HMS *Lutine* was carrying a huge cargo of gold and silver to Hamburg when it was blown onto sandbanks off the Dutch coast, and wrecked. All but one of the crew perished and the cargo was lost. The Lloyd's underwriters

---

[155] Contrary to popular belief, Lloyd's is not an insurance company. It is an insurance market, providing facilities for many independent insurance syndicates to operate.

who had insured the ship settled the claim in full, to the benefit of the reputation of the market. Efforts made to recover the precious cargo were unsuccess-
ful, but the ship's bell was salvaged. It was brought back to Lloyd's, then based
in the Royal Exchange building, and was hung in the underwriting room. From
that time on, when news arrived at Lloyd's of the safe completion of an insured
ship's voyage, the Lutine Bell was rung twice. When the news was of the loss of
a ship, it was rung only once.

Lloyd's is famed as the place to find an underwriter to cover any insurance
risk you might think of. When Tom Jones wanted his chest hair insured, or
David Beckham his legs, Lloyd's would offer a price. The details of the risk
would be written down on a slip, and taken around the underwriting room.
Once a lead underwriter[156] had been found to commit their syndicate to bearing some of the risk, and a suitable price had been agreed, others would follow.
As ever, sharing the risk is of the essence when it comes to insurance.

Nowadays, Lloyd's of London occupies an iconic building designed by architect Richard Rogers in the City of London. Perhaps fittingly, this very modern
building retains its original Victorian frontage, a reminder of its past. It remains
a marketplace for obtaining insurance of every kind. The Lutine Bell, now
gilded, hangs in a place of prominence in the modern underwriting room.

I use the word 'average' a lot throughout this book. It's time to look at its origins
in the shipping business, and how it came to have the meaning it commonly has
today.

average (n.)

(late 15c.), 'any small charge over freight cost, payable by owners of goods
to the master of a ship for his care of the goods', also 'financial loss
incurred through damage to goods in transit', from

avarie (French) 'damage to ship' and *avaria* (Italian): words from
12c. Mediterranean maritime trade. Of uncertain origin, sometimes
traced to:

awariya (Arabic) 'damaged merchandise'.

Thus, when for the safety of a ship in distress any destruction of property is
incurred, either by cutting away the masts, throwing goods overboard, or in

---

[156] The pricing expert who assessed the risk and considered it worth insuring would write his
name on the page under the details of the insurance. That person had 'underwritten' the insurance,
and so became the underwriter.

other ways, all persons who have goods on board or property in the ship (or the insurers) contribute to the loss according to their average, that is, according to the proportionate value of the goods of each on board.

Century Dictionary

This meaning developed to 'an equal sharing of loss by the interested parties'. The transferred sense of 'statement of a medial estimate, proportionate distribution of inequality among all' is first recorded in 1735. The mathematical sense 'a mean proportion arrived at by arithmetical calculation' is from 1755. The sports sense, as in a batting average, is first found by 1845, originally in reference to cricket.[157]

London's burning

When fire swept through the city of London in September 1666, Samuel Pepys's house was in the path of the blaze. In his diary entry for 4 September, he records:

> Sir W. Pen and I to Tower-streete, and there met the fire burning three or four doors beyond Mr. Howell's, whose goods, poor man, his trayes, and dishes, shovells, &c., were flung all along Tower-street in the kennels, and people working therewith from one end to the other; the fire coming on in that narrow streete, on both sides, with infinite fury. Sir W. Batten not knowing how to remove his wine, did dig a pit in the garden, and laid it in there; and I took the opportunity of laying all the papers of my office that I could not otherwise dispose of. And in the evening Sir W. Pen and I did dig another, and put our wine in it; and I my Parmazan cheese, as well as my wine and some other things.

That night, 13,000 homes (but only six lives) were lost. Fire is frightening and capricious. In Pepys's London, with timber-frame houses, narrow streets, and open flames used for light and heat, a fire could start in one of your neighbours' houses and imperil yours. That Great Fire of London in 1666 was the event that provided the impetus behind commercial fire insurance in the United Kingdom. In the aftermath of the fire, when Christopher Wren drew up plans for a new street plan for London, he explicitly included a site for 'The Insurance Office'. The plans were never carried out, but when London's first fire insurer was formed a short time after, it was indeed named 'The Insurance Office'.

---

[157] As ever, I am indebted to etymonline.com for this etymology and further expansion.

From the start, fire insurance was as much about prevention of damage as compensation for loss. The Insurance Office arranged fire brigades; these teams would rush to burning buildings and, if they saw the company's firemark (a metal badge above the front door certifying that it was insured), they would do what they could to limit the damage (and hence the claim payout). Soon, competing insurers were founded, each with their own brigade, which sometimes resulted in the absurd scene of a team of firefighters arriving at a fire, but refusing to intervene as the building was insured with a rival company. In time, the brigades started to collaborate and on the first of January 1833 were brought under control of a central body, the London Fire Engine Establishment.

Fire evokes an almost primitive terror for us, one that exemplifies a risk worth insuring against: when a fire takes hold it is shocking, seemingly random, and can be devastatingly costly. Fire insurance policies usually include stringent conditions for fire prevention and safety measures, and before large fire claims are paid out, thorough investigations as to cause will be carried out. With fire insurance in particular, though, there is ample opportunity for fraudulent claims: the moral hazard that a fire will be deliberately set as a way to realize the value of an unwanted building.

Life and other contingencies

Life and Other Contingencies is the curious title of a (now-superseded) textbook for actuarial students that taught the intricate calculations that life insurance actuaries need. The variety of risks that can be covered by insurance policies (and the intricacy of the calculations needed) astonished me as a student. A 'whole of life' policy was perhaps the simplest: a policy that would pay out a benefit on death, whenever it might happen. But even that kind of policy had its variations: for example, it could be paid for by a single initial premium, or by a periodic monthly or annual premium. If the policy did not continue for life but expired without value after a number of years, it was a 'term' policy, while if it paid the sum insured at the end of a fixed period, it was an endowment assurance. A policy that paid out an income for life was a life annuity, which might or might not have a guaranteed period (it looks bad if an annuity pays out just once or twice before the annuitant dies).

But the policy that most intrigued me was the tontine. Under this scheme, a small group of members agree to each contribute an initial lump sum, which buys each of them a regular annuity. Then over the course of years, as each of the participants dies, their share is distributed to the surviving members of the group, whose incomes accordingly rise each time one of their fellows dies. Finally, the last one standing is in receipt of all the income. It has not escaped the attention of crime writers that tontines, or similar last-survivor arrangements, provide ready motives for tales of murder. The moral hazard involved, the temptation to see off your fellows, was not restricted to fiction, and in consequence, tontines are no longer offered as legal contracts. Nowadays, life insurance regulations will only permit life assurance to be contracted where there is an insurable interest: you can't use insurance to make what is in effect a bet to profit from the death of another.

A fair premium

The risks affecting the lives, the motorcars, the houses, and other valuables that an insurer covers are far more complicated to measure than the probabilities associated with cards, dice, and wheels in the casino, but they're governed by the same principles of probability. But how can insurers calculate the probabilities on which they base their premium rates?

The world around us is messy. Every time you start your car there is some chance you will be involved in a collision, but the probability of that is not like the probability of rolling a pair of dice at a craps table. Thousands of factors are at play: the roads you're driving on, the weather conditions, the condition and qualities of your vehicle, your own driving skills. It might be easier if every driver had the same skills, drove the same model of car, and made exactly equivalent journeys every day. But we are all different. It's an example of the **uniformity-variability** duality. The insurance underwriter would love a uniform population to analyse, but is faced with variability. They must find some way of boiling down all the relevant aspects of the risk into a manageable set of data that can reasonably be captured in an insurance application form. Based on the wider claims experience, the underwriter must then create a model to calculate the approximate risk, relating it to this data. The premium that is charged for a policy should fairly reflect the expected cost of claims the insurer might have to bear.[158] The probability of claims is important, but the size of possible claims is, too. Insurance for a Lamborghini costs more than for a Ford Focus.

---

[158] This pure risk premium is not the only consideration. The insurer will also be thinking about many other commercial considerations, such as their competitive position, perhaps pricing more keenly than justified in order to gain market share.

Suppose that Jason, a 22 year old living in North London, is looking for a quotation for motor insurance for a 2008 Ford Fiesta. Jason is a competent driver, and has no history of previous claims or driving convictions. Even though there's nothing exceptional about Jason, it is not possible simply to calculate the average claims cost for similar policies: it's unlikely that *any* policies on the insurer's books will *exactly* match Jason's case. Rather, the pricing model must treat the make and model, the place the vehicle is kept, the age and driving history of the driver as separate factors. The pricing actuary will look for correlation between the available rating factors, and the patterns observed in the costs of claims arising from previous experience. They will then use statistical techniques to build a useful pricing model from the imperfect data available. The aim will be to provide reasonable terms on which to offer the insurance: not so expensive as to be uncompetitive, not so cheap as to incur losses for the company.

Broadly speaking, most non-life[159] insurance policies are priced so as to reflect the risk and costs of claims in the year immediately ahead, plus allowance for expenses and, of course, for some profit. This is not the case for life insurance. If a whole-of-life policy were to be fairly priced, year by year, then the premium to be charged should increase by approximately 10 per cent every year, as the risk of death increases by about 10 per cent for every year of age. However, the practice with life insurance is to offer a level premium. So for a young person, the premiums paid for such a policy will exceed the year-by-year risk, but this will in time be offset by premiums far lower than the true risk when they reach older ages. This is, on balance, a good thing. From the policyholder's point of view, the level premiums mean that continuing with the insurance does not become a burden late in life. From the insurer's point of view, it locks in the policyholder, providing an incentive for them to continue paying the premium for the policy in what is an increasingly good deal. In return for the policyholder's commitment, the insurer commits to provide cover in future years, without reassessing the risk to them.[160] This also allows the insurer to build up very substantial funds from the excess paid early in life, to be released late in life, but which in the interim stand to the insurer's credit. These reserves held against future claims turn life insurance companies into major investors and give their fund managers great power in the financial markets.

---

[159] It's never satisfactory to define something by what it's not. But non-life is a term in common use for those insurances other than life and pensions. It's also referred to as general insurance or, in the United States, casualty insurance. None of these terms works very well.

[160] This is why this kind of insurance is sometimes referred to as 'permanent'.

Insurance pricing actuaries make careful analyses to understand the risks they are covering and to correlate the risk with data they collect from policy applications. We know that correlation is not causation. Jason might have to pay a hefty premium for his car insurance, because he is a 22 year old male,[161] and so is a member of a group that is statistically associated with risky driving and high claim costs. But a statistical association based on a group does not necessarily apply to every individual member of that group. Jason may in fact be a very careful driver. While age and gender may be strongly associated with certain driving styles, they are nonetheless only proxies that allow Jason to be allocated to a category, rather than factors that define his driving habits, and Jason may with some justification feel that this is unfair. It's the **individual–collective** duality again.

Recently, in the case of car insurance, a different approach has become available. With the advent of car telematics and mobile communication, new and potentially more useful rating factors have become available. A box of electronics fitted to your car, or even a mobile phone app, can gather data in a way that allows a crude assessment of your actual driving behaviour: the suddenness of your stops and starts, your patterns of speed and acceleration, how sharply you make your turns. Monthly mileage and the overnight location of the car can be established as a matter of fact, rather than taken on trust. The exact circumstances surrounding an accident can be much more clearly understood through tracking data and accelerometers.

All of this means that proxies such as gender and age can become unnecessary. Risky drivers can be asked to pay a higher premium because their driving is objectively risky, and not just because they belong to groups that historically have shown risky behaviour. The super-cautious Jason can be judged on his actual driving behaviour once the telematic data starts to provide real evidence. Less cautious Tracey, when provided with proof of her erratic stops and starts, might learn to be a more careful driver if her insurance is based on telematics, and she can earn a lower premium for improved driving. This approach may be good for Jason, Tracey, and their insurance carrier, but it does have implications for privacy. Jason might be unwise to use his car as the getaway vehicle in a bank robbery if his insurer is tracking his every journey.

Too much information?

Skilful underwriting means accurate assessment of risk. But the principle of insurance is precisely that victims of bad luck should not suffer the full financial

---

[161] Motor premium rating based on gender is no longer permitted in the European Union.

impact of that bad luck. To protect the pool, the premium must reflect the risk that each policy brings. If too much information is known, then the bad risks may find themselves excluded from the pool, or allowed entry only on unafford-able terms. Once risks are too precisely known, the pooling effect may be less useful. If you had perfect knowledge of future claims, insurance would function as mere saving, the premium simply funding the known future payouts.

Analysing chance events in the real world always means looking at muddy, messy measurements and trying to identify the factors that could explain the variation, until we are left only with pure randomness. In recent years, geneti-cists have started to decode the messages hidden in human DNA. We under-stand more and more of how our genetic information accounts for our health through life, and we need to choose how to use this understanding. If, for example, an individual's DNA suggests a very strong possibility that in midlife they will develop Huntington's chorea, then an insurer would have some justi-fication in treating that potential almost as a pre-existing condition. What then happens to the pooling effect of insurance? If someone brings a greater risk to the pool, it seems only fair to expect them to pay a higher premium for that reason. On the other hand, it seems unfair that someone born with 'bad DNA' should be condemned not only to a life of poor health, but also expensive healthcare or expensive health insurance.

Reinsurance: You're going to need a bigger pool

Insurance operates on the principle of pooling of risks. If the financial blow of loss due to fire, theft, disease, or death can be shared among thousands of others, then by paying an affordable premium, individuals can sleep easy at night. But what about the actuary concerned with the financial wellbeing of the insurance company? What pillow cushions her head when she considers the prospect of a hurricane, or a towering inferno, or an exploding oil rig, with the potential for multiple claims, no longer independent, but coming together in battalions? Well, she in turn will have taken out an insurance policy to protect the funds under her care. The arrangements that she has made with reinsurance companies will allow her to make claims in the event of a catastrophic event, or for an out-of-the-ordinary pattern of losses. Cushioned by this reinsurance, the insurer is able to offer more extensive policies than would otherwise be possible.

Reinsurance companies also benefit from having data gathered from all their clients, the direct insurers. That means they can pool not only the risks, but the information relating to the risks. Their statisticians can use this bigger data set to generate better analyses and develop a better understanding of risk. Reinsurers become risk specialists, able to provide pricing (and cover) for risks that direct insurers would have difficulty managing.

And the reinsurance actuary will sleep more soundly knowing that he too, has his own reinsurance arrangements in place for still further pooling of the risks, relying on the fact that the capacity of a huge global marketplace can so dilute the impact of extraordinary claims as to allow even the worst disaster to be survived, at least financially.

It's entirely possible that as one reassurance company insures another, the chain of reassurance will form a complete circle, and the reinsurance under-writer may end up covering some of their own losses. As a result of the loss by fire of the oil platform *Piper Alpha* in 1988, it was discovered that such a reinsur-ance spiral did in fact exist, and some reinsurers found themselves liable for a second round of claims, as via indirect routes they had, in part, reinsured them-selves. And not just for a second time: some Lloyd's syndicates found they had insured some proportion of *Piper Alpha* many times over.

Central to the idea of insurance is that the risks covered are at least to some degree independent, and that means the law of large numbers can operate, so that the insurer bears an averaged risk. When the risk becomes systematic, though, when multiple claims arise from the same cause, whether due to spirals in the reinsurance market, or large-scale flooding, then larger and larger pools are needed to absorb the claims. For global events such as the potential effects of climate change, there may be no reinsurance treaty capacious enough to cover the losses.

Holding back the flood

After serious flooding in the United Kingdom in 2000 resulted in high levels of insurance claims, there was a risk that some homes built on water meadows and other areas at high risk of flood would become uninsurable. An agreement was struck between the UK Government and the insurance industry whereby the insurers would agree to continue to provide cover at then-current rates, and the State would invest heavily in flood defences. For various reasons this arrange-ment became unworkable, and in 2013 the agreement was discontinued in favour of a new scheme based around a dedicated flood reinsurance company, Flood Re. The scheme applies only to homes built before 2009 (to avoid provid-ing an incentive to build new homes on flood-prone land) and allows insurers to offer flood cover to their customers at a price that does *not* fully reflect the true risk of flood, but is to some degree subsidized. The costs are absorbed by Flood Re, which is funded by levies from the insurance industry without reflecting the specific risks of individual properties. The scheme is intended to operate for 25 years, regarded as sufficient time to put in place adequate flood defence measures. Finally, the government will act as an insurer of last resort, in the event of a flood of a magnitude that might be expected once in every 200 years. Given a changing climate, who can now be sure of how often we might experience what would have been called a 200-year flood? When risks are not independent but arise from catastrophic events affecting many people and properties at once, and driven by systematic underlying change of conditions, even the biggest of pools may not be sufficient.

Answer to:
What's the worst that could happen?

The quick answer is that there is *no* amount that would be a sufficient stake to back such a proposition. When you underwrite a bet like that, you take on an unlimited liability.

This tale is a variation on what is usually called the St Petersburg paradox. In its usual form, it suggests that you are offered the following deal. A coin is tossed. If it comes up heads, you win $1. If it comes up tails, it is flipped again, and if it then comes up heads on the second throw, you win $2. If it takes three throws for the first head to show you win $4, and so it continues. The longer it takes for the head to show, the bigger the prize becomes, doubling with every successive throw.

The question is: how much should you be prepared to pay in order to play this game with its guaranteed winnings? You can reckon the average expected winnings in this way. You have half a chance at winning on the first throw, so that opportunity must be worth half of a dollar: 50c. Then you have the chance of winning on the second throw. That is a 1 in 4 chance, and the prize is $2. A 25 per cent chance of $2 is also worth 50c. The chance of winning on the third throw is 1 in 8, with a prize of $4: again, this is worth 50c. And so it goes: every successive flip is an opportunity to win, and each offers half the chance to win twice the amount. Every one of these opportunities works out to be worth 50c. Since there is no limit to the number of opportunities available to win, the value of all these chances taken together must be an infinite amount of money.

That's the logic of the matter. But this seems to violate common sense: you'd probably be reluctant to offer even $10 to play this game. After all, it's very unlikely that you will get even that much back. This is the ludic fallacy at work: the arithmetic of probability derived from games can't always be carried over directly to real situations. The resolution of this paradox lies not in the arithmetic, which is sound, but in practical and psychological factors. A win on the twentieth flip would mean a prize of over half a million dollars: it would be fair to ask if that debt could even be honoured. And the value we place on a windfall may not be proportionate to the size of the win: the mathematics of the paradox relies on the idea that a minuscule chance of a massive win is equivalent to a half chance of winning one dollar. Arithmetically this may be so: psychologically they are very different. Ask anyone who regularly plays the lottery.

CHAPTER 17

Shaping the Risk

The winds and the waves are always on the side of the ablest navigators.

Edward Gibbon

Double or quit?

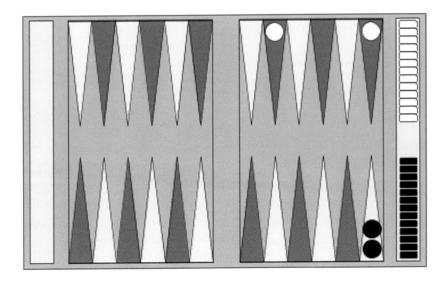

The final stage of the board game backgammon is a race to remove the playing pieces ('stones') from the board, using the numbers rolled by a pair of dice. In the position shown, each player has already removed 13 pieces from the board. The pieces are travelling to the right, as shown on the diagram.

What are the Chances of That? How to think about uncertainty. Andrew C. A. Elliott,
Oxford University Press. © Andrew C. A. Elliott 2021. DOI: 10.1093/oso/9780198869023.003.0017

It's White's turn to play, with one stone that is five spaces ('points') from completion and one stone that is just one point away. If White rolls any combination of numbers that includes a 5 or a 6, both pieces can be taken off, giving White the win. Further, since in backgammon double rolls are counted as four times the number shown, White will also be able to win if the roll is a double 2, a double 3, or a double 4 (but a double 1 will not be enough). If White fails to win in this turn, though, Black will certainly win, as any roll will permit the removal of the two black pieces.

What are the chances that White will succeed?

Backgammon is a gambling game, played for money or points. The rules of the game allow White, before rolling, to offer Black a 'double'. If Black accepts, the stakes will be doubled. If Black refuses the double, the game is forfeited to White, at the original, undoubled stakes.

In this situation, should White offer a double?

If White does offer the double, should Black accept, or forfeit the game to White?

Answers at the end of the chapter.

Making your own luck

(On it being suggested that he was a very lucky golfer)

Yes, I am, and the more I practise, the luckier I get.

Gary Player

Akiko Yazawa is a recent world champion backgammon player, and the first woman to hold that title. She won in 2018, having taken the title previously in 2014. For a game where dice play such an important role, the lists of backgammon champions do not suggest a random process is at work. For example, the women's champion in every year from 2011 to 2015 was the aptly named Laura Monaco (the venue for the championship is Monte Carlo). Where does this consistency come from, when the moves available to each player are entirely dependent on the numbers they roll? Backgammon is not like chess or *Go*, which are self-evidently games of skill, and where the available moves are entirely determined by the board position. In backgammon, if the roll of the dice goes against you, even if you are the world's best player, you will not win that game. If the luck is with you, there's little your opponent can do: victory is yours.

And yet, almost paradoxically, backgammon is without question a game where skill dominates over luck. A great backgammon player is a great backgammon player, and they will win a disproportionate number of times. Matvey 'Falafel' Natanzon, who died in February 2020, had taught himself to play, and had progressed from having to sleep in Manhattan's Washington Square Park to becoming widely acclaimed as the world's greatest player.[162] What made him so good? Certainly not luck. There are other factors that make backgammon winners, and understanding those factors can throw some light on strategies for managing chance in other areas of life.

For a start, a backgammon match is not just a single game. The winner of the Monte Carlo final is the first to 25 points (a simple victory in a single game earns the winner a single point, but very strong wins (known as gammons and backgammons), or the use of rules which allow stakes to be doubled, can increase the points won. But generally, to earn 25 points requires multiple games to be played and won. This allows the law of large numbers to come into effect, and the random impact of luck to be smoothed out a little.

Then, the way in which chance enters the game is important. In games such as craps, the dice roll is decisive and marks the end of an element of play. In backgammon, though, the dice roll occurs at the start of the turn, and offers multiple options to the player. How the players make use of the luck they receive is where skill comes in.

Finally, and possibly most important, a good backgammon player will make plays with an eye to the future, to prepare the board for future rolls. With two dice, 21 distinct rolls are possible.[163] A good position is one in which as many as possible of those 21 rolls offer good moves that can further improve the position. Some might call them lucky rolls, but the opportunities for luck will have arisen from skilful play in earlier turns. Conversely, a good board position is also one that has relatively little potential for unlucky rolls. Then, besides seeking to improve their own position, a skilled player will also seek to place the opponent in an unfavourable position: one with more chances for unlucky rolls and fewer lucky ones.

---

[162] Falafel Natazon was the commentator in the 2018 match that Akiko Yazawa won. The match was recorded and is available on YouTube (https://www.youtube.com/watch?v=ygO3JkTaIwk). It's worth a watch, even if only to understand the interplay of luck and skill involved.

[163] In backgammon the rolls (2, 1) and (1, 2) cannot be told apart and so are not distinct. Nonetheless, of course, a roll consisting of a 1 and a 2 is twice as likely as, say, a double 1, since there are two ways it can happen.

So, while the game is ostensibly a race to remove pieces from the board, in reality for much of the game it's about manipulating probability distributions in order to make that race easier. The state of the race is assessed by looking not only at the simple measure of how close to the finish line the stones are (the 'pip count'), but at the shapes those stones make on the board.

Akiko Yazawa has survived cancer. When it was diagnosed, she analysed her position in terms appropriate for a backgammon player: 'When you think about your next move, you picture all of the possible outcomes. So I pictured the best outcome of surgery, and the worst. And then I pictured the best outcome of not getting the surgery, and the worst... not undergoing the surgery was not an option.'

There's no escaping the role of chance in a backgammon game, and no hope of predicting or influencing the roll of the dice. But you can prepare for the different ways in which the game can develop. Single stones on the backgammon board ('blots') are vulnerable, and a cautious player may choose to limit their exposure. Doubled stones are strong, and when they form a block, they can limit your opponent's options. Still other shapes of the stones offer potential for building strong positions in future turns. In the face of uncertainty, preparation is sometimes more helpful than prediction. Limit your exposure to adverse chance, maximize your opportunities to exploit good fortune, and minimize the consequences of bad luck.

Hedging your bets

hedge (v.)

> late 14c., 'make a hedge', also 'surround with a barricade or palisade'; from hedge (n.).
>
> The intransitive sense of 'dodge, evade, avoid committing oneself' is first recorded in the 1590s, on the notion of hiding as if in a hedge.
>
> That of 'insure oneself against loss', as in a bet, by playing something on the other side is from the 1670s.

To hedge a bet is to offset that bet with its opposite. Hedge funds got their name from the strategy of playing both sides of the market, and they have become a symbol for risky activity in financial markets, speculation for speculation's sake. It's something of an irony, then, that derivatives, the types of investment most

associated with hedge funds, were originally intended for controlling and limiting risk.

Those who produce the commodities we consume are exposed to various
risks. Among the risks that a corn farmer in the US Midwest faces is uncertainty
over the price his crop will fetch. Each year Farmer Jones makes an investment
in planting his crop, and in tending to it as it grows. The costs of producing the
current harvest are sunk. When, in due season, he sells his produce, if he relies
on the open market, the price that his goods will command is uncertain. It may
be that he is unlucky, that when he comes to sell, market forces will have driven
the price down, so that the return he achieves is less than the price he needs.
That would hit his profits. On the other hand, he could benefit if the price was
unexpectedly high. On the whole, though, gambling on the price might be a
risk he would rather not take.

Financial engineering to the rescue. One of the choices available to Farmer
Jones is to sell his crop in advance, using a 'futures' contract. The contract is a
promise to deliver a certain quantity of corn on a nominated day, for which he
receives a fixed price. Now his downside price risk is removed. Of course, he
has removed the upside too: by agreeing the price in advance, he can no longer
profit from a higher price, but, he may well reason, his business is farming, not
speculation.

Today, in mid-February 2020, you can trade in corn futures in the Chicago
Board of Trade Market at a price of $380 for a contract covering 5,000 bushels
of corn (around 2.25 metric tonnes), to be delivered in March 2020. (For
comparison, the same amount of corn for immediate delivery is priced at
around $395, while for delivery in May 2020 the price is $384). So, Farmer
Jones can sell a futures contract, and fix his price of $380 today, against the
promise of delivering 5,000 bushels of corn next month. If the price is higher
next month, too bad, he's missed the chance of a better price. If the price has
crashed, he has secured his price.

But this contract has itself exposed Farmer Jones to another risk. He is now
obliged to deliver the contracted corn on that date. If his crop fails, his obligation
still remains, and he may, perversely, be required to buy corn elsewhere in order
to fulfil the contract, and if corn is in short supply, that may be costly. He has
shifted his risk, but he has not eliminated all risk.

There are other strategies available to him, though. He can use an option
contract. This is a contract that gives him, as the name suggests, an option, a
choice to sell his harvest at the contracted price. This seems like a good plan.
For one thing, if his crop fails, he has no obligation to make use of the option,

he can simply tear it up. For another, if the price on the day is higher than the price stipulated on the option, he needn't use the option at all, he can simply sell at the higher market price and take the greater profit.

There is, of course, a price to pay. The option contract itself costs money, and whether he uses it or not, the farmer will have borne the cost of that contract. It's much like an insurance policy. If you need it, you'll be very happy to have paid for it. If you don't, you'll get nothing back, apart from the peace of mind that comes from knowing your risk was covered.

Today, in February 2020, Farmer Jones can pay $150 for an option to sell a consignment of corn (a 'put' option[164]) at a price of $380 next month.[165] If, when the date arrives, the price is higher than $380, he can simply sell for whatever price he can get. If the price is lower than $380, though, he can exercise his option and protect the value of his crop. This protection comes at a cost: $150 is a substantial bite out of his profits. But the figure of $380 is not the only 'strike price' available. He may prefer instead to buy a cheaper option that guarantees him a lower price, say $370. That option, which is not worth as much to him since it protects a somewhat lower price, costs only around $20.

When the option strike date arrives, the value of the option will differ, depending on what the market price on that date is. Here's a chart showing the value to Farmer Jones of his put option at a strike price of $370:

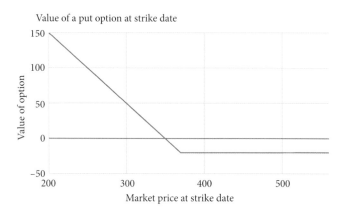

Value of a put option at strike date

Value of option

Market price at strike date

If the price at the contract delivery date is $350, then he breaks even: he uses the option to sell his corn for $370, but the option has cost him $20, and so he

<hr />

[164] Options to sell are 'puts', options to buy are 'calls'.

[165] There is a distinction between 'American' options, which can be exercised at any point before the expiration date (or on that date), and 'European' options which can only be exercised on the expiration date.

nets $350. He is in the same position as if he had taken the market price. If the price is higher than that, he has lost his $20 option price for no gain. If the price is below $350, though, buying the option contract will have been worthwhile.

Willingly accepting greater risk

You don't have to be a farmer to be interested in corn prices. Futures and options were created to reduce the risk relating to the underlying commodity (in our example, corn) but they can have a life of their own. An option is like a playing card in a game; it gives you the choice to make a certain move. It has a value of its own, and can be bought and sold between players in its own right. The only link to the underlying commodity is that changes in the price of that commodity will be the main influence on its own price.

If a professional trader (let's call her Trader Barnes) were to buy a put option for the ZCH20 contract (the Chicago Board of Trade code for corn to be delivered in March 2020) at a strike price of $370, the profile of her potential profit from that deal would be identical to that of Farmer Jones.

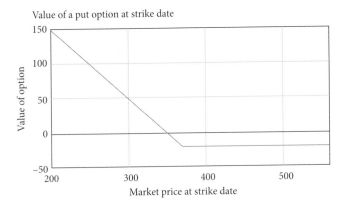

Value of a put option at strike date

But while Farmer Jones will have done his deal to protect himself from adverse price fluctuations, aiming to reduce his risk, Trader Barnes is buying her option contract in order to bet that the price will be low, perhaps anticipating that there will be a surplus of corn on the market in March. She is betting $20, and if she is right, if the price is less than $350, she will win her bet. The put option that she considers buying is exactly the same as the one that Farmer Jones buys, but while he buys it to reduce risk, she buys it to take on a risk.

Trader Barnes could make other kinds of bets on the price of corn. A *call* option is an option to *buy* corn at a specific strike price. Here is what the profit/loss profile for a call option to buy corn futures in March 2020 at a price of $390 looks like:

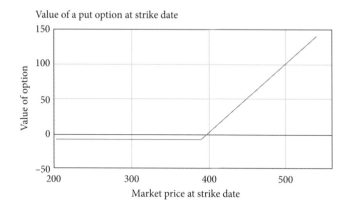

If the price of corn is above $400 at the strike date, Trader Barnes will make money on this bet, since she has the option to buy it at a cheaper price. Both this call option and the put option shown are 'out of the money': they would not be worth exercising at the current price.

These are some simple examples. Greater complexity comes when you start to combine options. If you think of these contracts as building blocks, traders like Barnes will be able to construct compound bets, or 'strategies', of all kinds. Here is what the risk profile looks like for her if she decides to buy one of each of those options, betting both that the price will rise and that it will fall, but losing money if the price stays close to where it is now. If she adopts this strategy, she is betting on volatility and against price stability.

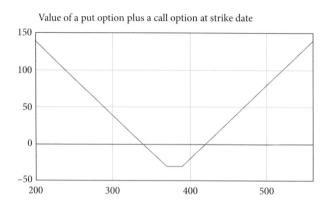

As you can imagine, there are many possible variations: Trader Barnes can choose different strike dates for her contracts; she can *sell* option contracts as well as buy them (when she does this, her profit may be limited but her losses could potentially be large). Contracts of this kind give financial players with an appetite for risk a way of investing that is in effect pure risk.

As the strike date approaches, the value of the option will change, until at the strike date it is either zero, or close to the difference between market price and the strike price. It might be worth nothing; it might be worth a great deal more than the price originally paid. The price risk inherent in the corn price has been distilled, made more potent. There is no fundamental value in an option; everything is in the movement of prices: pure upside and downside.

As things turned out, the price of corn at the strike date of 13 March 2020 was just a little under $370. Farmer Jones's option contract to sell at $370 would have been worth something, but not as much as the $20 he would have paid for it.

Getting technical

Using these building blocks, traders can shape their exposure to risk arising from the underlying investment in an astonishing variety of ways:

Trader Barnes's combination of an out of the money (OTM) put and call as shown above is termed a *strangle*. If the put and call were both 'at the money' (ATM), the shape of the profit profile would be a 'V' and would be called a *straddle*.

A *reverse iron butterfly* is constructed with two puts (one ATM and one OTM) and two calls (one ATM and one OTM). It looks like this:

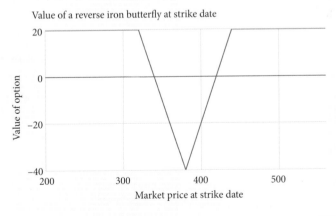

Value of a reverse iron butterfly at strike date

continued

> ### Getting technical *Continued*
>
> With this strategy, a modest and limited profit is achieved if there is a large enough shift from current prices, while giving a loss if the price doesn't change much. You might guess that an *iron butterfly* strategy would look similar, but inverted, profit for loss. You'd be right. That would be a bet on price stability.
>
> Through the right combination of puts and calls, you could construct many different shapes. What you cannot freely choose is the positioning of the profit profile on the vertical axis. There must be a balance between potential profits and losses, the level of that balance determined by the markets.
>
> It can get more complex than this. These option strategies are all combinations of options expiring on the same date. By involving options of different dates, Trader Barnes can bring into the picture the way the value of an option changes as it approaches the expiry date.

These contracts, futures and options, are termed derivatives, as they are derived from the underlying assets,[166] and they are an essential part of responsibly managing investment portfolios. They can be used to shape a portfolio's risk profile in a sophisticated way, by covering vulnerabilities and enabling opportunities.

The value of assets held in other countries will, in addition to the normal commercial risk, also be vulnerable to a currency risk. If you were an American investor invested in the shares of an Australian bank, you might see their value fall not only if the bank does badly commercially, but also if the Australian dollar falls in value against the US dollar. If you'd prefer not to bear that currency risk, you could make use of options for trading AUD against USD so that any losses on the Australian bank shares that were caused by currency fluctuations could be offset by matching gains on your options.

The range of derivatives and their combinations may make you think of the various combination bets available in horse racing, or the options presented at the craps table, where a single source of randomness is used as the basis of any number of derivative complex bets. These are ways in which gamblers and traders alike are able to shape their exposure to risk.

Every option has its price

In 1967 Ed Thorp (he of the card-counting and roulette wheel exploits) published a book called *Beat the Market*. He had devised a way to calculate the

---

[166] The name appropriately also recalls the 'derivative' from calculus, the rate of change of a function.

prices for a type of derivative called a warrant, which entitles the holder to buy shares at a given price. He had seen that these were sometimes overpriced in the financial markets. His system exploited this anomaly and involved selling warrants and buying the underlying stock in precise ratios to create a strategy that would be profitable regardless of market movements. By 1969 he had generalized this approach and was actively trading options using his new pricing formula. He kept the formula secret, though. In his words:

> I had this options formula, this tool that nobody else had, and I felt an obligation [to the investors in his hedge fund] to basically be quiet about it. The tool was just an internal formula that was known to me and a few other people that I employed.

Four years later, two academics, Fisher Black and Myron Scholes, published 'The Pricing of Options and Corporate Liabilities', a paper in which they independently derived formulas for option pricing based on the same approach (and in which they acknowledged Thorp's earlier, but more limited work on warrants). They also give credit in their paper to the work of Robert Merton. Their option pricing formula is now generally known as Black–Scholes, or sometimes Black–Scholes–Merton.

The price of an option depends on the strike price, and the current price, and the time to expiry, but, critically, also depends on the *volatility* of the price of underlying security, or commodity. Volatility is a measure of the uncertainty of price movements. If prices varied little, options would have little value, but the greater the uncertainty, the greater the value and hence the price of options.

The availability of a reliable derivative pricing calculation drove a huge expansion in the market for derivatives. The original model has been widely criticized, though, as it assumes that price movements follow a normal distribution. This assumption of normality is now known to be incorrect, since market prices are more prone to extreme movements than the normal distribution allows for. Nonetheless, adaptations of the formula are still widely used.

Systematic risk

Any investment is associated with a variety of different types of risk. In the 1990s nothing could have seemed safer than investing in bank shares, but they, too, have been proven to be risky. In the first place, there is the risk associated with the choice of one particular bank. Perhaps Natbarc plc will make bad choices, and so perform worse than other banks. Perhaps its decisions will be smart, so it performs better. By deciding to invest in Natbarc, you've exposed yourself to the risk implicit in that choice.

There is an obvious strategy to reduce that risk. You don't have to invest in just one bank: you could invest a smaller amount in each of several banks. Then, if one bank does poorly and another relatively well, things even out. By diversifying, you smooth the bumps a little. But you are still exposed to the kind of market sector risk that affects all banks equally. Tough conditions may mean hard times for all banks. By choosing to invest in the banking sector generally, you lessen the risk of a bad choice of a specific bank, but retain the risk associated with the banking sector generally.

So you can diversify your portfolio to include other kinds of financial firms—insurers, maybe, or property companies. You're still faced with any risks that might have an impact on financial services companies generally. And so it goes, level after level. The wider the range of securities you take into your portfolio, the more the large numbers will help you, and so the more the risk (and of course potential for reward) is diluted. Diversifying reduces risk, but only to the extent that the constituent parts are independent of one another.

In 2008 the world economy was shocked when several large banks were brought down by a financial crisis, and several others had to be rescued by governments. An important factor behind that crash was the failure of the law of large numbers. Or rather, the unjustified reliance that financial engineers placed on the law of large numbers. They were trying to pull the trick that casinos and insurers use, of making risky business safe through the averaging that happens when you're dealing with large numbers. The thinking went like this:

If a company lends money to an individual to buy a home, that becomes a liability for the borrower (they have a debt that must one day be repaid), but an asset for the lender (they will one day be repaid). From the lender's point of view, though, it is a risky asset. Repayment is not absolutely certain—the borrower may go bankrupt, or for some other reason default and never repay the loan. Lenders can and do assess borrowers' capacity to repay: those they rate highly get favourably low interest rates; those who seem more risky will be labelled with the horribly euphemistic term 'subprime' and will have to pay a higher rate of interest. But if the rate of interest is set at a level that is higher than usual to reflect the degree of risk, then, taken over a large enough portfolio, things will average out (so the thinking goes) and the additional interest paid by all the subprime borrowers will compensate for those random few who do not repay the loan. Individually, the loans are risky: collectively (in theory), an invest-ment based on a collection of those loans is supposedly far less risky. In the early 2000s, subprime mortgages were widely available in the United States, in some cases more or less regardless of the borrower's creditworthiness. After all, most borrowers would find a way to keep up the payments (who wants to lose their home?) and, ultimately, the loans would be backed by the security of the actual property.

Now things get a little more complicated. These assets, these entitlements to repayment that the lender has on the books, can be passed on to others. Take a large collection of these contracts, arrange them into tiers based on credit-worthiness, bundle them together, and offer the bundles for sale to investors. If the credit rating agencies buy your logic that by collecting together a large number of loans, even the riskiest ones, you have managed to make a low-risk bundle, then you have spun gold from straw. On paper, you are selling an asset with a high rate of interest, where the underlying risk of default has been aver-aged away: a small proportion of defaults is expected, but the numbers are large enough for this to be offset by a higher interest rate. The investment bank can then offer participation in these bundles, a slice of the profitable action, to its customers.

All of this would be perfectly in order if the underlying risks (of borrowers being unable to repay) were truly independent of one another, or even only moderately correlated. If financial hardship struck borrowers entirely randomly, then there would be some justification in claiming that the risk had been diluted by the bundling together, that the law of large numbers would work its magic.

They were wrong. When the economy goes sour, jobs are lost, not independ-ently, but in waves. Many people are hit at the same time. In the 2008 crisis, the rate of default was huge. Worse, in assessing the risks for each bundle of loans, the ratings agencies had relied on average measures (such as the average credit scores of the lenders), but had not taken into account the way in which those measures varied within the bundles—they fell foul of the **uniformity–variabil-ity** duality. To make things even worse, the financial crisis meant that house prices collapsed—so even when the lenders repossessed the properties that sup-posedly provided security for the loans, they ended up with large numbers of homes that were worth far less than they had been, and in some cases were in effect valueless.

These defaults started the collapse of a house of cards. Ever more complex financial structures had been built on the base of the subprime home loans and the financial bundles that had been created from them. And because the banking system formed a tightly connected network, trouble in one bank could easily and rapidly spread to others. One bank after another found itself in grave difficulty. And at the root of all this was that the risk assessment had assumed only a weak correlation between mortgage failures. The law of large numbers depends on the independent behaviour of the individual parts, and they had misjudged the likelihood of a systemic effect on their complex investments.

Answer to:
Double or quit?

White is about to roll, and to win must roll a combination that allows their two remaining stones, on the 1 point and on the 5 point, to be removed from the board. There are 36 possible rolls, and the ones that win for White are anything with a 6 or a 5, plus double 4, double 3, and double 2. Count those up, and you find there are 23 winning rolls. White has a 23 in 36 chance of winning.

Assume the stake is $1. White can offer to double those stakes. If White chooses not to offer the double, the game plays out at the original stake of $1. With 23 chances to win against 13 chances to lose, White has a net advantage of 10 chances in 36, and that is worth $$^{10}/_{36}$, approximately $0.28.

If White offers the double, there are two possibilities: Black may refuse, in which case White wins $1 immediately. If Black accepts the double, White has the chance to win $2, and now the net advantage will be twice as large: $0.56. Both are better outcomes, so the best choice for White is to offer the double.

What about Black? The offer having been made, is it better to concede and lose $1, or to allow White to make their final roll at twice the stakes, and risk losing $2? Black weighs a certain loss of $1 against an average expected loss of $0.56. It makes sense for Black to accept the double and retain a $^{13}/_{36}$ chance of winning, at doubled stakes, if White fails.

I've left out some of the rules that affect the calculation of the odds in backgammon, but none that would change this simple endgame situation. For example, if one player wins before the other player has taken *any* pieces off, it scores as a double win and is called a *gammon*. A win where the other player still has pieces in the opponent's home board (the most remote quarter) is a *backgammon*, and is scored at triple value. This means that sometimes the position for the leading

SHAPING THE RISK | 313

player is *too good* to offer a double: the opponent would refuse, and the player with the advantage would lose the chance for a multiple-point win.

If a double is offered and accepted earlier in the game, the accepting player gains the right to redouble, and in a volatile game you may see this right (or rather, its physical representation, the doubling cube, with faces marked: 2, 4, 8, 16, 32, and 64) being passed from player to player as the stakes increase.

Good backgammon play requires a keen sense of how these factors interact, and how to manage the chances and risks that emerge through the game.

CHAPTER 18

What's Coming Next?

Science has not yet mastered prophecy. We predict too much for the next year and yet far too little for the next ten.

Neil Armstrong

The Prophetess and the Prince

A long time ago there was a city ruled by a wise King. The King's son noticed that whenever the old man made his decisions, he placed a great deal of trust in one particular adviser, an old woman they called the Prophetess. When the Prophetess warned of disease in the city, the King announced strict isolation and quarantines. When the Prophetess warned of famine, the King ordered rations cut. When the Prophetess warned of war, the King ordered the soldiers to tend to their weapons and to spend many hours training. But as far as the Prince could see, the predictions that the Prophetess made were always wrong. The city was never swept by disease; the people never starved; the enemy never attacked. To the young Prince, that Prophetess seemed to do nothing but spoil everybody's pleasure, and all for naught.

In time the old King died, and the Prince was crowned. Some months later, as the winter was ending, the Prophetess came to him saying there were signs of a new disease in the city: the young King should immediately shut down the taverns and order a curfew. He listened to her politely and then made up his own mind. The taverns would stay open. In the weeks that followed, an infection took hold and hundreds of people died.

What are the Chances of That? How to think about uncertainty. Andrew C. A. Elliott,
Oxford University Press. © Andrew C. A. Elliott 2021. DOI: 10.1093/oso/9780198869023.003.0018

In the summer, the Prophetess came to the young King and said she feared the harvest would be a poor one. He should consider rationing the stocks of food in the granaries. But the young man had planned a great festival: how could he cut rations for the people and still have his feast? That winter many of his subjects starved.

As the next spring came around, the old woman came to him and told him that his enemies were preparing for war. Their city was weak: he should instruct all able-bodied people to strengthen the walls and prepare to fight. This time the young King listened, but the disease and the famine had taken their toll. The preparations for war were insufficient: the walls were breached and the city was taken.

As they were led away in chains, he asked her: 'How is it that when you advised my father, you were always wrong with your warnings, but with me you have been right every time?'

I'm sure you can guess her answer.

Should I take an umbrella?

There are at least two good reasons to use statistics to try to build coherent models of the world from what might at first seem like random observations. The first reason is backward-looking: to seek explanations of what has happened. A century ago there was a surge in cases of lung cancer, and no one knew the cause. Now we understand why. Statistical analysis has allowed us to make coherent models that explain both the patterns and the randomness of diseases and other aspects of our uncertain world.

But the second reason for building meaningful models is to look forwards. A good model can explain the past, but a more rigorous test of a model is its ability to make useful predictions in new situations. Courts of law may concern themselves with resolving the uncertainties of past events, but for most of us, the burden of uncertainty means worrying about what might happen in the future. We want a crystal ball.

Fitzroy: Easterly 3 to 5, backing south-easterly 5 or 6

On 1 August 1861, *The Times* in London published its first ever weather forecast for the day ahead. The prediction (covering the United Kingdom divided into three areas) was:

North—Moderate westerly wind; fine
West—Moderate south-westerly; fine
South—Fresh westerly; fine

The forecast had been provided by Robert FitzRoy, a person of huge energy and achievement. He had been commander of the *Beagle*, the ship on which Charles Darwin had taken his influential voyage from 1831 to 1836. FitzRoy would go on to be elected, in 1841, as the member of the United Kingdom parliament for the Durham constituency. Shortly after that, he was appointed as the second governor of New Zealand, in which post he served from 1843 to 1845. He returned to the United Kingdom in 1848 and retired in 1850, due to ill health. Then in 1851, with Darwin's support, he was admitted to the Royal Society—then, as now, an organization of immense scientific prestige.

In 1854, though, FitzRoy and a staff of three started work in a new organization, the Meteorological Department of the Board of Trade, that would in time become the United Kingdom's Meteorological Office. Joseph Henry had in 1831 demonstrated the use of the telegraph as a way of communicating over distances, and over the succeeding two decades it had become one of the great innovations of the day. For the first time, the telegraph allowed information to be gathered from widely dispersed locations and rapidly communicated to a central point, and this made it possible for FitzRoy to form a coherent picture of the weather conditions in many different places at a single point in time. Taking advantage of the telegraph by setting up networks for gathering weather information, FitzRoy used this information to generate and publish short-range weather forecasts. He also invented and distributed new types of barometers, recognizing the connection between air pressure and short-term weather prospects. A fall in atmospheric pressure is a good leading indicator of imminent bad weather.

At first, FitzRoy's weather forecasts were received positively. After all, the very concept of systematically predicting the weather seemed miraculous. But the novelty soon wore off. Inevitably, FitzRoy's predictions were sometimes wrong and, despite their undoubted value, those who came to rely on his forecasts started to become critical when they were inaccurate. The *Cork Examiner* reported: 'Yesterday, at two o'clock, we received by telegraph Admiral FitzRoy's signal of a southerly gale. The gallant meteorologist might have sent it by post, as the gale had commenced the day before and concluded fully twelve hours before the receipt of the warning.'

In April 1865 Vice-Admiral FitzRoy, wounded by criticism, troubled by depression, and beset by financial difficulties (having spent his fortune on developing weather forecasting infrastructure) killed himself by cutting his own throat with a razor. The weather forecaster is an easy target for mockery, and he had come to believe that his efforts at forecasting had been a failure. Today, though, he is regarded as a pioneer of meteorology. One of the United Kingdom's recognized sea areas is now called FitzRoy: his name is recited and remembered every day when a radio announcer reads out the BBC's shipping forecast.

Forecasting is a tricky business: perfect accuracy is not possible. It's been estimated that FitzRoy's storm warnings were right about 50 per cent of the time, and their availability must have saved thousands of lives.

Although short-range forecasts are now quite accurate, long-range weather forecasts remain much less so. The fundamental reason for this is obvious: forecasting involves piling uncertainty upon uncertainty. The weather for tomorrow depends on the weather today. Tomorrow's weather forecast can use today's measurements, which may be accurate, but the forecast for the next day must use tomorrow's numbers, which are themselves uncertain. And so, day by day, each forecast compounds uncertainty to the point where daily detailed prediction is impossible. And that piling of uncertainty upon uncertainty is characteristic of systems, like the weather, that are called *chaotic*.

Chaos

Chaos: When the present determines the future, but the approximate present does not approximately determine the future.

Edward Lorenz

This iterative nature of weather forecasting—the idea that each new state of the weather system is dependent on the one before—creates a particular challenge. It's not just that the weather is complicated and that the calculations are onerous and extensive. The deeper problem is that weather systems have inherent mathematical features that make them chaotic. Used in this technical way, *chaotic* refers to processes where the way in which things develop is extremely sensitive to the precise starting conditions, usually because of feedback effects.

When Edward Lorenz was running computer simulations of the weather in 1961, he wanted to reproduce results from a previous run of a computer program. To save time, he typed in numbers from a printout of the numbers representing an intermediate state of the earlier calculation. However, when he ran his program, he found that the results were very different from those he had seen previously. He finally found the discrepancy: the numbers he had typed in from the printout had been rounded to three decimal places, whereas the computer's internal storage had used six decimal places of accuracy. These very small differences—of the order of one in a thousand—resulted in very different results in the iterated calculations. Where some might have dismissed this effect as an annoyance, Lorenz investigated it, and he is now regarded as one of the founders of chaos theory, the theory of such systems: deterministic, and yet for practical purposes unpredictable.

Chaos theory explains how some processes that follow completely rigid rules can nevertheless give rise to outcomes that are wildly unpredictable and seem quite unstable, even random. The 'butterfly effect' whimsically asserts that a small flutter of air caused by a butterfly flapping its wings in Amazonia can, through propagation and amplification of atmospheric disturbances, result in tumultuous storms over Europe. Attempts to predict the state of chaotic systems quickly illustrate the impossibility of precise forecasting in the long term: the smallest of deviations can produce wildly different outcomes.[167]

Well, should I take an umbrella or not?

When I look at the BBC weather app on my phone, it gives me a percentage likelihood of rain for every hour of the day. What does this percentage mean? Clearly a bigger number means more chance of rain, but what does the percentage really represent? Here is the explanation provided by Nikki Berry, meteorologist at MeteoGroup, which provides the forecasts for the BBC:

> As counter-intuitive as it might seem, hourly predictions of the chance of rain actually refer to the 60 minutes preceding—not following—the time mentioned...if you imagine putting a measuring jug out in your garden at 13:00, and you go and check it at 14:00, the 40% chance of rain at 14:00 is the chance of you having some rain in your jug at 14:00. If you're heading out at 14:00, looking ahead to the forecast at 15:00 will give you some guidance for how the weather might play out.

How are those percentages arrived at? Using a technique called ensemble forecasting, the meteorologists run many simulations of the weather system, using data based on the measured observations, but including a little random adjust-

---

[167] Not all systems are chaotic. Many systems are stable, and even self-correcting, converging in predictable ways. This is the sort of behaviour that Lorenz might have expected when he restarted his weather calculations from a halfway position.

ment, a little dither in the data. Since the models are chaotic, small variations in the data give a wide range of outcomes. The weather forecasters can then count how many of their simulations result in, say, rain in my location, and turn this into a probability of rain tomorrow, or the day after, or next week. They monitor their accuracy by checking how often they get it right. Of all the times they predicted a 40 per cent chance of rain, they count the cases where rain did indeed fall, and measure themselves by how close this was to 40 per cent of the time. Through better measurements, ensemble forecasting (using vast computing power), and refinement through monitoring their success, weather forecasts have steadily improved over the years. The United Kingdom's Met Office now claims that its four-day predictions are as accurate as its one-day predictions were 40 years ago.

Watching the political weather

February 2020: snapshots in time

The next section tracks a historical process—the selection of the Democratic Party candidate for the 2020 United States presidential election—as it played out. By now, you will know how history looks back on this process. But at the time of writing each entry in this diary, I did not. Everything lay ahead.

There is a mental shift that happens when the uncertain future becomes the established past. The eventual outcome shapes the narrative of our memory of the process. We easily forget the uncertainty of ignorance when with hindsight we can read the history of what actually happened. It's useful to highlight this **foresight–hindsight** duality by reflecting on the uncertainties as they developed.

As I write this in February 2020, the primary season in the United States is about to start: caucusing and voting to select the Democratic nominee to run in the 2020 general election.[168] Analysts at the website FiveThirtyEight.com have built a forecasting model to analyse the chances of each candidate hoping to win the nomination. Nate Silver, FiveThirtyEight's editor-in-chief and statistical expert, reports that building this model has been more challenging than any they have previously tackled (this is the first time they've attempted a

---

[168] There seems to be no serious challenge to Donald Trump as the Republican Party's nominee.

detailed forecast for the presidential primaries). The main reason for this diffi-culty is that this is an extended process that is strung out over months, as opposed to the kinds of election they usually model, which take place on a single day/night. The chain-like nature of this primary process means that earlier stages can (and will) influence later stages. Who does well and who does badly in the Iowa caucuses (happening today, as I write) will influence the modelled probabilities for later stages. More convoluted feedback is involved too: the early results will influence the strategy of the participants, and this will change what happens in later stages (and this in turn will be reflected in changes to the model parameters). Some candidates will drop out on bad results, and their supporters may transfer their allegiance elsewhere. This chain-like nature means that the model has the potential to be unstable. Nate Silver characterizes the model as 'path-dependent and non-linear'.

3 February 2020—Eve of the Iowa caucus

Here's the assessment of the front-runners today, on the eve of the first big primary event, the Iowa caucus:[169]

Biden	Sanders	No one[170]	Warren	Buttigieg	All others
2 in 5	3 in 10	1 in 6	1 in 20	1 in 30	<1 in 100
43%	31%	17%	5%	4%	

5 February 2020—After the Iowa caucus

In the Iowa caucus the data-gathering technology that assembled the results from caucus locations was flaky, meaning that as of today, there is no final out-come. It seems likely, though, that Buttigieg has done best, with Sanders also making a good showing. FiveThirtyEight analysts have frozen their forecast: they had made no allowance in their model for the possibility of a delayed result. Media attention has now switched to the next contest, the New Hampshire primary, and the impact of the as-yet-unknown Iowa result may be diluted. So this is one of the reasons that forecasting is such a difficult practice:

[169] The numbers quoted represent the predicted probabilities for each candidate of getting a majority of delegates by the time of the Democratic convention—the convention process itself is not modelled.

[170] 'No one' refers to the scenario where no candidate gains a majority of delegates by the time of the party convention, which would make for some exciting politics at that time.

the only basis on which we can make credible models is to utilize data and patterns from the past, but a 'black swan'[171] event such as this failure of technology in Iowa can break the model.

Update (later the same day): the forecast model has been updated. We now have:

Sanders	No one	Biden	Warren	Buttigieg	All others
2 in 5	1 in 4	1 in 5	1 in 10	1 in 20	<1 in 100
37%	27%	21%	10%	6%	

The numbers have changed dramatically. Why? What has happened here? Biden, Sanders, Warren, and Buttigieg are the same people as on Monday morning. Their policies and characters have not changed, and there have been no revelations of scandal. What has caused the probabilities to change? It's due to the availability of new information. What was previously only supposed (that Buttigieg and Sanders would do well in Iowa, and Biden poorly) is now confirmed. The Biden–Sanders rivalry has been reversed. But despite the success of Pete Buttigieg (his modelled chances have gone from 4 per cent to 6 per cent), he remains in a weak position.

Should we say that the previous probabilities were wrong? Will the new probabilities themselves turn out to be wrong? What do these probabilities represent anyway? They cannot correspond to a frequentist interpretation of probability: the 2020 Democratic primary is a one-off event. It cannot be replayed over and over. FiveThirtyEight says that Sanders now has a 2 in 5 chance of reaching the convention with a majority of delegates; but this process will not be played out five times, still less in the numbers required to establish statistical credibility. What will be played through multiple times is the model that Nate Silver has built. Typically, this model is run 10,000 times and the probabilities quoted are drawn from those results, in much the same way as the meteorologists use ensemble forecasting to give us a probability of tomorrow being a rainy day.

The model that lies behind these numbers will have captured FiveThirtyEight's understanding of the primary process and how the events of that process affect one another. For example, success for one candidate in a caucus or a primary election typically results in a 'bump', a boost in the opinion polls, which in turn

---

[171] The term 'black swan' was coined by Nassim Nicholas Taleb, and refers to a wholly unexpected event, never even considered as a possibility.

may help their chances in subsequent contests. Part of the analysts's model concerns how large this effect may be. Then too, the states are each different from one another. Iowa's electorate is not the same as New Hampshire's, and the model aims to reflect these distinctions (and similarities, where states are similar). It incorporates not only the chains of cause and effect, but also quantifies the strength of these links by using statistical associations drawn from past experience.

11 February 2020—Eve of New Hampshire primary

Sanders	No one	Biden	Warren	Buttigieg	Bloomberg	All others
2 in 5	3 in 10	1 in 6	1 in 20	1 in 25	1 in 40	<1 in 100
44%	28%	16%	5%	4%	3%	

Sanders's position is now looking increasingly strong, while Biden has slipped to a very weak position. Michael Bloomberg has attracted a lot of attention: his chances are now rated highly enough to place him among the front-runners.

12 February 2020—After New Hampshire primary

Sanders	No one	Biden	Buttigieg	Bloomberg	Warren	All Others
2 in 5	1 in 3	1 in 6	1 in 20	1 in 30	1 in 30	<1 in 100
37%	34%	17%	5%	4%	3%	

In the New Hampshire primary, the story of the night was the so-called Klobucharge—a surge in support for Amy Klobuchar. And yet she does not appear on the table. And perhaps that shows one of the virtues of a model like this: enthusiasm for Klobuchar made a good story for television, but setting the need for narrative aside, the story that the numbers tell is that it made very little difference to her prospects. But maybe the model is wrong. Time will tell.

1 March 2020—After South Carolina primary

No one	Sanders	Biden	Bloomberg	Warren	Buttigieg	All others
3 in 5	3 in 10	1 in 9	1 in 100	<1 in 100	<1 in 100	<1 in 100
51%	28%	11%	0.6%	0.1%	0.1%	

'No one' has a substantial lead! It seems possible that this could lead to a contested convention with no definitive front-runner when the Democratic convention comes around in July 2020. Bloomberg's run is starting to look like a washout. Warren and Buttigieg's chances have slumped: they appear to be effectively out of the race now.

3 March 2020—Dawn of 'Super Tuesday'

No one	Biden	Sanders	Bloomberg	All Others
2 in 3	1 in 5	1 in 6	<1 in 100	<1 in 100
63%	21%	16%	0.1%	

Biden is now the front-runner candidate, although 'No one' retains the top spot by a huge margin. Sanders has slipped. All others have vanishingly small chances. It's now a two-person race.

6 March 2020—After 'Super Tuesday'

Biden	No one	Sanders	All others
7 in 8	1 in 10	1 in 30	<1 in 100
87%	10%	3%	

An astonishingly good performance by Biden on Super Tuesday leaves him as the presumptive nominee. Sanders is still in the race, but with 3 per cent chance, the model offers him slim hopes.

Is that the end of the story?

4 April 2020

It's now almost a month later. The coronavirus crisis has erupted in the past month and made talk of presidential primaries seem a very secondary matter: another reminder that even the best models rely on things progressing, in some measure, as they have in the past. The Democratic Party's nomination process is stalled. However, to all intents and purposes, the contest seems to be over. FiveThirtyEight's model has Biden as >99 per cent likely to enter the convention with a majority of delegates.

Biden	No one	Sanders
>99 in 100	<1 in 100	<1 in 100
>99%	0.2%	0%

8 April 2020

Sanders has withdrawn from the race. Joe Biden will contest the election as the nominee of the Democratic Party.

So what?

What is the point of all this? Did this model allow us to predict the future? No. Were the initial predictions a good guide as to what would happen? Not really, if you look at the way the lead flipped from Biden to Sanders and then back again. Did the model's calculated probabilities represent any sort of objective truth? Emphatically, no.

So what's the point, then? Does the model allow us at least to make better bets on the political prediction markets? Possibly, yes. The odds offered on prediction markets more or less tracked this rollercoaster ride, but not exactly. To the extent that the model reflected an unbiased analysis, it might have given insight as to when the betting markets offered odds that were objectively too high or too low. But even this possible advantage is somewhat tentative.

What the model did offer was a rational and more or less dispassionate way of incorporating the developments as they happened, a way of tracking the twists and turns that was better than mere punditry. When Pete Buttigieg dropped out, the model allowed FiveThirtyEight staffers to write an article headlined 'Buttigieg dropping out isn't good for Sanders'. Using the model, they could explore the consequences of events in an unbiased way[172] that would have been difficult for commentators relying only on their political nous and personal judgement. When Klobuchar 'charged' in New Hampshire, it excited a lot of pundits. Not so much the FiveThirtyEight bunch, who saw that her chances had indeed risen, but only to 0.2 per cent—scarcely worth talking about.

Predictions in other fields can of course be more accurate than these political calculations. Astronomers predicted the transit of Venus in 1639 by analysing data and building models of the movements of celestial bodies—and the model

---

[172] Or at least biased only to the extent that the model's parameters incorporated any deeper bias.

they had constructed was sufficiently accurate and remained sufficiently stable to allow good predictions to be made. The calculations were validated when observers saw the silhouette of Venus start to track across the face of the Sun, right on time.

The end is nigh (once again)

When it comes to human affairs, we are all too ready to predict the worst. Prognosticators of doom always have an audience: it seems that each successive generation becomes convinced that the world is falling apart and is headed for disaster, and it's very easy to make a prediction that if things continue to deteriorate in the way that they have been, we will all come to a sticky end.

Thomas Robert Malthus

The name of Malthus has entered the English language:

Malthusian (adj.)

> of or relating to the theories of T. R. Malthus, which state that population tends to increase faster, at a geometrical ratio, than the means of subsistence, which increases at an arithmetical ratio, and that this will result in an inadequate supply of the goods supporting life unless war, famine, or disease reduces the population or the increase of population is checked.

The definition above summarizes the idea for which Malthus is best known, initially expressed in his *Essay on the Principle of Population*, published in 1798. He argues that the human population grows through an exponential process— that population tends to grow by the same *proportion* every year, or decade, or whatever time period you choose. In contrast, food production is a linear growth process: he held that food production could only ever increase at a steady rate, by a constant *amount* in each successive period. So according to Malthus, in the long run, population growth would always outstrip food supply: the number of people would always grow to eliminate any surplus in food, leading inevitably to shortages.

This idea is a gloomy one, leading to one of two natural consequences. Either the population is held in check by famine, disease, and war, or moral restraint may be needed, to limit both reproduction and consumption. Given human nature, the former must seem more likely.

Were Malthus's ideas right? There are two planks to his argument: exponential growth of populations and linear growth of resources. Are they valid? Malthus was living at a time when the population of the world was around 1 billion. He thought that population would, given sufficient resources, increase exponentially. And through most of the twentieth century that was more or less true. The world's population was growing at least exponentially, the rate of growth peaking at around 2 per cent per annum in the 1970s. But since then, the world population growth rate has come down, and not because there are great numbers of people dying from famine, disease, and war. The model has changed. Current UN population forecasts show that the world's population is now growing at around 1 per cent a year, and the total population size is expected to stabilize towards the end of the twenty-first century at a level of around 11 billion.

The second plank of Malthus's argument was that agricultural production would increase only in a linear way. It's not easy to discern his reasoning here, but it seems to boil down to this: 'I can't imagine how you could squeeze more food out of this fixed amount of land: maybe every twenty-five years you could add as much capacity as we currently have, but you could never achieve exponential growth'. Malthus can hardly be blamed for failing to anticipate the successive revolutions in efficiency of food production that have led to food now being cheaper in real terms than ever before. Mechanization, intensification, use of chemical fertilizers and insecticides, and higher-yielding crop varieties have meant that the cost of production of food has reduced dramatically.[173] Global trade means that food can be grown in those places where it is most efficient to grow it. A wealthy nation which buys in a significant proportion of its food is effectively paying to bring more land (in other countries) under agricultural production for its benefit.

Malthus thought that this imbalance meant that population numbers could be stabilized only through misery or restraint, but the stabilization in population size that is now happening is associated with voluntary limits on family size, arising from better options for birth control, empowerment of women, and relative prosperity. If Malthus could look at the world today, he would see that the model he formulated has changed in ways he could not have anticipated.

---

[173] This is not to say that these innovations have not themselves caused problems and been costly in other ways. But that wasn't Malthus's argument.

Projection based on the continuation of the same underlying model may give useful results in the short term, but when we look at longer time scales, unanticipated changes to the rules of the game mean that simple mathematical extrapolation is unreliable.

Malthus's ideas were convincing at the time and remain superficially seductive. It is easy to see the logic of projecting outcomes based on a continuation of the current trajectory. Part of that logic is that unchecked natural growth is exponential and, in the terminology of our age, unsustainable, when compared to known and limited resource availability.

The limits to growth

The Club of Rome, founded in 1968, describes itself as 'an organisation of individuals who share a common concern for the future of humanity and strive to make a difference'. In 1972 the Club published a book called *The Limits to Growth*, a highly ambitious project that used computer modelling to make global economic forecasts for the decades ahead. A team of 16 researchers, working under Dennis Meadows, constructed a very elaborate model (called World3) of the interactions among parts of the global economy and the available resources, with a distinct Malthusian flavour.[174] The conclusion was that the world was headed for imminent disaster. In the report's 'standard model' projection, world population was predicted to increase rapidly until around the middle of the twenty-first century and then collapse due to shortages in food and medicine. Many variants of the model were proposed in an attempt to examine the effect of different kinds of intervention.

A model of such ambition and complexity, seeking to represent not just gross effects but involving intricate chains of cause and consequence, can get things very wrong if those causes and consequences are not modelled correctly. The more parameters the model incorporates, the more places there are for errors to creep in and multiply through feedback effects. Any model with strong feedback effects will, like models of weather systems, be vulnerable to small variations in the input parameters.

The book and follow-up reports remain controversial, with some maintaining that, in broad outline, the issues raised by the book remain of concern.

---

[174] A key idea of the model was that resource usage increased exponentially and not linearly, meaning that finite resources would be exhausted much more rapidly than had previously been anticipated.

Nonetheless, the more testable of the book's conclusions, for example that world economic production per capita would peak in around 2008, have not been borne out.

The Population Bomb

In 1968 Paul Ehrlich, professor of biology at Stanford University, published *The Population Bomb*, a book warning of imminent (in the 1970s and 1980s) mass starvation, in all parts of the world. In his lifetime he had seen the world's population double from around two billion to four billion, and he argued that a further redoubling was under way. Whereas Malthus could see prospects for limited growth in resources, Ehrlich considered that we were pretty much at full stretch in terms of feeding the world.

Of course, there was no mass starvation in the seventies and eighties. Food calories per person have increased since Ehrlich wrote his book. He was right in that the population will soon have doubled again to reach 8 billion, but the rate of growth has fallen dramatically, and based on current projections will stabilize at around 11 billion.

As did Malthus, Ehrlich relied on projections based on a continuation of then-current trends. He could not imagine what could be done to boost food production, writing 'I don't see how India could possibly feed two hundred million more people by 1980.' By 1980, India's population had grown by around 170 million: by now, 2020, it has grown by a further 600 million. The Green Revolution transformed the picture in a way that few people could have foreseen. It's easy to be a pessimist, and see the problems accumulating. It's much harder to anticipate the innovative, unexpected solutions that can germinate in the minds of smart and motivated people.

How do you refute an argument that runs: 'we can't carry on growing the population like this—sooner or later we will run out of stuff'? Here is one possible counterargument: that our experience so far has been that human ingenuity and hard work have given us technologies that amplify our capabilities to such an extent that, far from becoming impoverished, the world's people (taken as a whole) have become more prosperous and overall living standards have improved. But for this argument to hold into the future, we are required to believe that ingenuity, which by its nature seems serendipitous, can indeed reliably deliver step-change improvements in the future, and this without destroying fragile natural systems. The evidence of the past suggests that this may be possible, but who is prepared to rely on chance? We'd have to believe that there are geniuses and visionaries—statistical outliers—among us right now. It's hard to place much faith in a policy of 'something will turn up', even if that's often how it's happened in the past.

There are some unfortunate consequences of such failed forecasts. On one hand, despite the failure of the predictions, the negativity associated with prognostications of doom seeps into the popular mood. A message of 'we're all doomed' leaves little room for optimism and little hope for progress, and can encourage a culture of nihilism, selfishness, and short-termism. Others will see the failure of forecasting and conclude that all predictions of disaster are flawed and can be disregarded. 'We've been here before', the worldly cynics declare, as climate change is proclaimed to be a hoax and warnings of the impacts of trade wars are dismissed as scaremongering.

The ozone hole

Sometimes, though, we do pay attention to predictions of calamity and take appropriate action. An example is the response to the ozone hole that developed over the South Pole and widened through the twentieth century. Alarming reports in the 1980s declared the existence of an 'ozone hole', a depletion of the amount of ozone in the stratosphere, centred on the South Pole. What had initially been dismissed as a measurement error was confirmed as a genuine phenomenon. Seasonal change meant that this hole would widen and contract on a yearly cycle, but year upon year, the situation was getting worse.

Since stratospheric ozone serves as a filter for damaging ultraviolet rays from the Sun, there were consequences for the ocean and for people living in places that were positioned under the widening hole, such as New Zealand and Australia, where the risk of skin cancer became significantly greater. Scientific opinion converged on a conclusion that a substantial cause of the widening ozone hole was the impact of chemicals used in daily life, notably the chlorofluorocarbons (CFCs) used in refrigeration and as propellants for aerosols.

In a rare example of international cooperation and coordination, the Montreal Protocol to ban production of CFCs was adopted in 1987, and it came into effect in 1989. The effects of the ban were noticeable by the mid-1990s, slowing the worsening trend, and by the end of the 2000s, a reversal was occurring. The biggest ozone hole ever recorded was that of the 2006 season, and since then it has been shrinking year by year. Recovery of the stratosphere has been slow, but based on current forecasts, it is expected that by 2075, ozone levels will have returned to pre-1980 levels.

The Millennium Bug

Some calamities are easier to predict than others. Perhaps the easiest of all is recognizing that a digital calendar that uses only two digits to record the year number will fail when a century boundary is crossed. In the 1990s it became clear that many computer systems that had been developed over earlier decades would, contrary to original expectations, still be in operation on 31 December 1999. These venerable programs, many of them essential for infrastructure, security, and commerce, often made use of a two-digit representation for the year. Bluntly, when the clock rolled around at the end of 1999, instead of starting the new year as 2000, many systems would treat the date as 1900. Calculations of duration or age would fail, being calculated as negative numbers. (Born in 1958, you must be –58 years old?) It took no statistical subtlety to correctly forecast disaster.

Nor did it take much insight to know the right corrective action to take: patch (and test!) the computer systems. At great cost and effort, thousands upon thousands of programs were amended, recompiled, tested, and successfully deployed. Not all systems were successfully corrected,[175] but there were no catastrophic failures, thanks to the diligent efforts of those responsible for fixing the systems.

A different narrative has now taken hold in the minds of much of the public. The Millennium Bug has become a byword for a failed prediction of disaster. This is deeply frustrating to those who committed time and resources to ensuring that disaster was averted. In the popular mind, the Millennium Bug is now regarded not as a catastrophe avoided, but as an unnecessary panic on the basis of an overblown warning. Dire consequences were warned of and were not in the end experienced, not because the warnings were wrong, but precisely because the warnings were heeded.

Knightian uncertainty and black swans

The Millennium Bug was entirely predictable. Even those programmers of the seventies, coding dates using two digits for the year, must have realized the compromise that they were making, saving two digits of storage for each date but incurring a technical debt to their future selves and their successors.

---

[175] Rather embarrassingly for them, the US Naval Observatory, that nation's chief timekeeper, showed the first date of the new year on its website as 1 January 19100.

But often the future has turned out to be different to what we expected, for reasons that were entirely unexpected. In 1921 economist Frank Knight wrote a paper in which he distinguished measurable risk from unquantifiable uncertainty. In Knight's view, if a risk could be quantified, it scarcely counted as a risk at all: it could be mitigated by methods based on statistics. Knightian uncertainty, though, was different. It was not to be found in the randomness associated with known trends or distributions, but in the arrival of something entirely outside the scope of what could be reasonably foreseen. This is not always bad news. Uncertainty brings **opportunity** as well as **disruption**, and Knight felt that uncertainty of this unmeasurable kind created opportunities for new businesses.

It's also easy to see that the failure of the predictions of doom made by Malthus, the Club of Rome, and Ehrlich could be explained by their understandable failure to anticipate technological change that would invalidate the model each had assumed. Nassim Nicholas Taleb calls extremely rare outlier events 'black swans'. They are so far outside of one's frame of thinking as to be beyond prediction. But he rejects Knightian uncertainty as being something categorically different from measurable risk, preferring the idea of a gradation of uncertainty 'at the level of the probability distribution itself'. When we make predictions, we do so on the basis of an underlying model, but that model itself may be uncertain, and arguably this uncertainty of the model may itself be quantifiable.

This still does not help us to make prophetic predictions. If some futures are beyond our ability to predict, do we then simply surrender to whatever chance might have in store? No—even if we cannot *predict* what is to come, we can *prepare* for the unexpected. We may not know when and how a pandemic may strike, but we can take reasonable measures to keep our health services and infrastructure in good order, ready for a crisis of whatever kind. We may not know when an earthquake will strike, but we can engineer earthquake-resistant buildings. We may not know exactly how our industrial activities are harming the natural world, but we can establish seed banks and conservation programmes and try to preserve habitats that maintain biodiversity.

Environmental destruction and extinctions

At various times throughout my life, from early childhood, I have been lucky enough periodically to visit the game reserves of Southern Africa, and see for myself the spectacular animals that belong on those grasslands and in those forests. On a recent visit, I was marvelling at the diversity of the distinctive coloration and patterning of the creatures: the waterbuck with a white ring around its rear end, the kudu with its pale wavy stripes down its sides, the spots of the cheetah, the fingerprint splodges of the leopard, the crazy-paving of the giraffe.

My first thought was one of wonder and the hope that these creatures would in the long run be able to avoid extinction. The quagga, looking like a half-finished zebra, is gone. Many other species are under threat of extinction. But another thought followed quickly. Even if we could protect and preserve all of these species, would the end result be anything more than a living museum? Three species of elephant are recognized: Asian, African bush, and African forest. In the absence of humanity's disruption, could there, now or in generations to come, be more than three species? Has the elephant stopped evolving? The impact of our dominance on this planet may be not only to drive countless species to extinction, but also, to put a stop to the further evolution of large creatures.

Large numbers are necessary for small chances to show themselves. We know that viruses and bacteria continue to evolve. So too do insects: they have the numbers. But where will we ever find the striped giraffe? The reticulated zebra? Existing populations, even if large enough to ensure survival of species, may be too limited to permit the continuation of the story of evolutionary diversification for these large creatures. Over billions of years, the Earth has been populated with animals of wonderful variety and diversity. Evolution needs many generations and substantial populations. I hope I am wrong. I hope that life on Earth does have the potential to continue this journey, and that our descendants, a million generations in the future, will have made space for sharing this planet with a new form of elephant, a different kind of panda, perhaps even a new species of great ape.

Nuclear war

In June 1987 a group calling themselves the Chicago Atomic Scientists published the latest edition of their group's *Bulletin of the Atomic Scientists*, which featured on its cover an image of a clock with the time set to seven minutes to midnight. This notional clock, now popularly known as the Doomsday Clock, was instituted as a means of conveying to the general public the extent of the threat ('the end of civilization') posed by the then-current global political situation in the light of the nuclear arms race. The 'minutes to midnight' measure has no absolute quantitative meaning: it signifies the group's opinion, and what matters is not the absolute setting, but the relative movements as the group decides to shift the time closer to or further from midnight. It is, in effect, a measure of subjective risk. In 1991, after extensive arms limitation talks, the setting was at 17 minutes to midnight, the most optimistic it has ever been.

The Doomsday Clock was devised to capture the severity of the existential threat facing humanity: risks that are so large that they cannot be smoothed away by the law of large numbers. In 1987 that meant nuclear war. But, starting in 2007, those setting the Doomsday Clock have started to recognize the threat posed to life on Earth not only by nuclear war, but also by extreme climate change. As I write this, the clock stands at its most alarming time ever, 2 minutes to midnight, tied with the setting in 1953, when the United States tested its first thermonuclear device. Although concern over President Trump's approach to nuclear arms control and negotiations forms part of this alarming assessment, the *Bulletin* leads with the threat posed by climate change. How can we form any sensible idea of the risk posed by that?

Climate change

We live with uncertainty in thousands of small ways every day of our lives. One way of containing risk is to play for small stakes. The gambler who wants to enjoy an evening in the casino should resist the urge to blow their budgeted betting money on a single spin of the wheel. Instead they should make smaller bets and so prolong the thrill. With climate change, though, there is no possibility of smaller bets. The stakes could hardly be higher, and when it comes to public debate, the waters could hardly be muddier.

When large companies are challenged in the marketplace by innovative start-ups, one of the weapons at their disposal is known as FUD—fear, uncertainty, and doubt. The established supplier simply has to make the argument that to entrust a large project to a relatively untried small supplier has significant, even if unspecified, risks. There is no need to bother with details: it can be enough simply to throw up a cloud of FUD. Those who wish to dispute the risk of climate change to the planet often use uncertainty in a similar way: as a weapon. In the seventies and eighties Big Tobacco did everything they could to obfuscate the science in regard to the growing body of evidence of the huge health impacts of cigarette smoking. And now the fossil fuel industries appear to be doing everything they can to delay the moment when the damage done through burning those fuels becomes undeniable.

WHAT'S COMING NEXT? | 335

Many of the risks that we have historically faced and survived have been overcome because those risks have been in some measure independent. John Graunt distinguished the 'chronical' diseases endemic in the population from the 'epidemical' ones, like the plague that swept through the population with devastating impact. Even so, populations survived large-scale risks like plagues and fires because enough people had been safe and saved. But the risk posed by climate change is of a different kind. There is no place to hide from it. No quarantine is possible. Risks cannot be averaged out when they affect everybody.

This is not the place to catalogue the overwhelming evidence for climate change. But it is a place to make a plea for understanding the nature of uncertainty. We now know that the risk of catastrophe arising from climate change is very substantial. But even if that risk was slight, the prudent approach would be not to dither, as governments seem to be doing, but to act, promptly and decisively. We have the technological capability to derive enough energy from sustainable sources. We have, collectively, the skills, wealth, and resources to feed the world, even when the population peaks at around 11 billion towards the end of this century. We have the capability to address this; to reduce and even eliminate the environmental emergency we face. But we continue to choose not to do these things. In many aspects of our individual lives we act with caution, sometimes excessive caution. But in this matter, in regard to our collective risk, we seem to lack the political will to act with proper regard to the risks we face, even to safeguard ourselves.

Last Chance

The most important questions of life are indeed, for the most part, really only problems of probability.

Pierre-Simon Laplace

Life is no simple game. There is no book of rules. But if we are smart, we can figure out some strategies that lead to better choices. We can interpret the clues that help us tell bad paths from good. Understanding the nature of chance and keeping a sense of proportion and perspective are essential to choosing well. But there are no guarantees of success, and even the canniest of players can find their strategy undone by a bad roll of the dice.

For millennia, people ascribed the effects of chance and the mysteries of uncertainty to the agency of deities and higher forces, often making literal sacrifices in an attempt to wrest some element of control back to the human realm. But over time we have developed the skills and built the intellectual tools to allow us to penetrate those mysteries and find clarity where before we saw only chaos. With understanding comes the opportunity for control. We have created medicines that prevent and cure disease, and have built safer means of transportation and better houses so that we are more likely to survive accidents and disasters, and we have set up public institutions that help to keep us safe from one another. We are moving, incrementally and unsteadily, towards a point where most avoidable deaths will be deaths that are in fact avoided. The great majority of us will die of causes that we can call 'old age'. This is progress. This is as it should be.

A sixth duality? Detachment–Engagement

When I started working on this book, I had a simple plan in mind: I would tell some stories about probability that were engaging enough to hold the reader's attention and informative enough to pass on some of my understanding of how chance works. I would dispassionately show what the science and mathematics of probability tells us.

But the more I thought about chance and the role it plays in the world, the more I began to try to puzzle out *why* thinking about chance is so hard, and why we fail to allow properly for uncertainty in the way we approach the future. This led me to the five dualities that I have articulated as themes in this book: individual–collective; randomness–meaning; foresight–hindsight; uniformity–variability; and disruption–opportunity. These five dualities are not mere intellectual puzzles to be relished for their Zen-like, quasi-contradictory nature. Very often they are what underlie the hard questions of chance and uncertainty that we face about issues that really matter.

Recently, governments in every country have had to make choices in their response to the Covid-19 pandemic. Some have argued that the cost of shutting down an economy was too great: others have held that the best way of protecting the economy is to protect people. These questions would be contentious enough to resolve even if all the facts were completely clear. They are made yet more difficult because these decisions must be taken against the background of huge uncertainty. These are very weighty matters and people rightly are attached to the ethical principles involved—and that makes a detached assessment of the uncertainties very difficult indeed.

And so this is perhaps a sixth duality to add to the five—one more reason that it is hard to think clearly about chance. Because for the most part we care deeply about matters where uncertainty is involved, and yet we think we should be dispassionate. Matters involving chance are very often matters of great importance. They very often involve us in personal ways, and very often concern issues about which we have strong prior beliefs. So our personal feelings must be reconciled with detached analysis.

The artist Georges Seurat was known especially for his pointillist painting technique. His paintings include sublime areas of colour that seem alive with shimmering textures. On close examination, we can see that these blocks are built up from tiny dots and dashes: tight, crisp brushstrokes of sharply intense colours. We experience our daily lives in this zoomed-in way—in moments of intensity.

But we can zoom out, detach ourselves a little, so that the view widens to include our neighbours' losing raffle tickets as well as our winning one: to encompass consideration of others' joys and sorrows alongside our own. And then the brushstrokes blur together a little and that shimmer appears, as those vibrant highlights add subtlety and life to the composition. We need to see the big picture, but should not forget that what brings the picture to life lies in the details.

In this book I have used ideas from games to explain matters of chance in ways that are simpler, and of less consequence, than their real-life counterparts. But life is no game, and the choices we make are not about simply winning or losing, but about the nature of what is won or lost. Each of us is dealt a hand of cards to play, and each of us has the opportunity to play it well or badly. Those of us lucky enough to have been dealt favourable cards have an obligation to use that luck well, to make the most of what we have and to try to give more than we take. Life is far from fair, but there are always opportunities to make it less unfair. Real choices have real consequences, and we can never be completely sure what those consequences will be. But I am sure of this: better information (more information, and more reliable) leads to better understanding, and better understanding of the nature of uncertainty leads (on average, and in the long run) to better choices.

Making your choices, taking your chances

A ship is safe in harbor, but that's not what ships are for.

William G. T. Shedd

There is one further meaning to the word chance:

opportunity (n.)

a time or set of circumstances that makes it possible to do something. From

opportunitie (late 14c.), 'fit, convenient, or seasonable time'. From

opportunite (Old French 13c.). From

opportunus (Latin) 'fit, convenient, suitable, favourable', from the phrase:

ob portum veniens 'coming toward a port', in reference to the wind. From

ob 'in front of; toward' and

portus 'harbour'

We all set sail on our own journeys, whether they end up hugging the coast, or crossing oceans. We all hold fears that our ship may be wrecked, and all have hopes that we will see the voyage through to our advantage, and finally will reach a quiet harbour in safety, eased in by an opportune wind. And along the way, as chance will have it, there will be incidents and accidents, adventures, triumphs, and defeats. Relish them all.

Life is good. Make it count. Take your chances.

SEE ALSO

This book is one among many that have been written about chance and uncertainty, and there are copious resources on the internet to help you learn more about probability and statistics. Here I have tried to collect a list of resources for additional reading that have stood out to me as being particularly interesting or useful.

General

Books

The Art of Statistics (David Spiegelman)
Do Dice Play God? (Ian Stewart)
Black Swan (Nassim Nicholas Taleb)

Websites

WhatAreThe.ChancesOfThat.com—The companion website for this book
UnderstandingUncertainty.org—Short essays and explanations of probability and statistics

Part 1: Pure Chance

Books

Roll the Bones: The History of Gambling (David G. Schwartz)

New Complete Guide to Gambling (John Scarne)
The Odds Against Me (John Scarne)
The Science of Conjecture: Evidence and Probability before Pascal (James Franklin)
Games, Gods and Gambling (Florence Nightingale David)
Entropy: God's Dice Game (Oded Kafri)
Entropy: A Guide for the Perplexed (Dwight F. Mix)

Part 2: Life Chances

Books

The Little Book of Luck (Richard Wiseman)
The Man in the High Castle (Philip K. Dick)
Observations etc. (John Graunt)

Websites

CambridgeCoincidences.org—Site for collecting coincidences
OurWorldInData.org—Highly recommended site for data to explain how the world is and why

Part 3: Happy Accidents

Books

The Beginning of Infinity (David Deutsch)
Gene (Siddhartha Mukherjee)
The Book of Humans (Adam Rutherford)
A Brief History of Everyone (Adam Rutherford)
The Ancestor's Tale (Richard Dawkins)
Climbing Mount Improbable (Richard Dawkins)
You Look Like a Thing and I Love You (Janelle Shane)

Part 4: Taking Charge of Chance

Books

A Brief History of Florence Nightingale (Hugh Small)
Smoking Kills: The Revolutionary Life of Richard Doll (Conrad Keating)
The Book of Why (Judea Pearl and Dana Mackenzie)
The Big Short (Michael Lewis)

ACKNOWLEDGEMENTS

Without the positive feedback I have received from readers of my previous book *Is That a Big Number?*, this book would not have happened. That support confirmed my belief that there is a place for a book that goes back to basics, re-examining how we think about numbers, without being overly technical, and showing how numeracy connects with every aspect of our lives. This book takes that project further: an examination of how we think about uncertainty. So my first thanks are to all those who bought and read the numbers book, and especially to those who took the trouble to express in words, to me personally, or publicly, their reactions to it.

Above all, though, I must thank Beverley, my wife, for her constant encouragement and tolerance. When I am writing, the chapter I am working on becomes a near-obsession and the smallest of setbacks or achievements is magnified. Throughout this process, Beverley has been my certainty amid uncertainty. Thanks are due, too, to my friends of many years who have listened with patience when I have bent their ear on some small or large topic that was occupying my brain. It is a great thing to have such friends, and especially over the last year.

I want to give particular thanks to Rose Malinaric, my sister. She has been my touchstone in writing this book. Her editing skills mean that when I send out even the earliest drafts to other readers, those drafts are in good shape. But Rose's influence is deeper than she knows: when writing, she is one of the readers I keep in mind, especially when it comes to shaping the language. Without her, this book would not exist.

Then too, I must thank Leslie Gardner, my agent, for her unfailing encouragement and enthusiasm. Whenever there has been a bump on the road, she has been there to sort things out. Thank you, Leslie.

Thanks also to Katherine Ward and Dan Taber at Oxford University Press for having confidence in this book, to Karen Francis, the copy-editor, and to Kim Stringer who compiled the index. The index of a book is seldom appreciated, but it's worth taking a look at what's been done here: it delights me. Thank you too to Sumintra Gaur and Manikandan Santhanam who have project-managed the production of the book.

In this book I tackle a topic that is difficult to grasp in an intuitive way: indeed, that is one of the chief reasons for writing it. I am not immune to that difficulty. While I have taken pains to ensure that my facts and figures are correct, I could never be absolutely certain of that. In fact, you might ask: what are the chances that somewhere in all these pages an error or two has slipped through? The answer, of course, is that it's a near certainty. I take responsibility for all of them.

INDEX